A Friendly Guide To
The Adolescent Brain

By Avery Bedows and Madeleine Bender

Illustrated by Madeleine Bender

With the Consultation of Dr. Eugene Lipov, M.D.

Avery Bedows

Contents

Taking Inventory of Your Brain

The human brain. An object so utterly complex that we have not yet been able to come anywhere close to replicating its extraordinary functionality. Your brain has somewhere between 100 and 200 billion neurons (cells that act as the information processors). This number, however, is dwarfed by the number of connections *between* neurons...by a factor of 100: it's estimated that there are 100 *trillion* of these connections in the human brain.

Similar to a computer chip, these individual processors are not piled together at random. The brain is organized into distinguishable structures, and while they're all interconnected, many regions serve specific purposes. The aim of this book is to delve into the aspects of neuroscience that are most pertinent to adolescents, but it is important that the brain's basic structure and function be addressed in order to put everything else in context. This chapter is brief and condensed and could not possibly be 100% retained after a single read, so you should refer back to it when necessary whilst reading other chapters.

Directions

The human brain is a complex three-dimensional structure. To help explain the location and orientation of its various parts, a system of directional labeling has become standard in the medical fields. Here are the basic directional indicators:

1. Front/behind – anterior/posterior
2. Away from/close to – distal/proximal
3. Near the top, towards the back/near the bottom, towards the belly – dorsal/ventral
4. Above/below – superior/inferior
5. Toward the side/toward the midline – lateral/medial
6. Toward the back – caudal

Bear these in mind, because they will come in handy if studying any level of anatomy.

Major Components

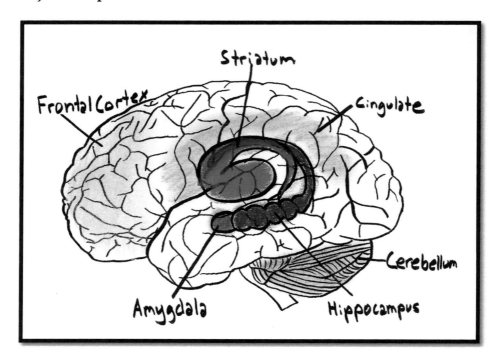

Neuroscientists have agreed on breaking the brain into three general parts: the **forebrain**, the **midbrain**, and the **hindbrain**. The forebrain contains the cerebral cortex, as well as some subcortical structures (the thalamus and hypothalamus), and is located in the *anterior* (see above) part of the brain. The midbrain is classified as the part of the brainstem that connects the forebrain and the hindbrain – which includes the brain stem and its components, as well as the cerebellum.

The Cortex

The **cortex** is the brain's outer layer, and is broken up into two halves, or hemispheres (the right and left sides of the brain). It has four lobes: the frontal lobes, the parietal lobes, the occipital lobes, and the temporal lobes.

The **frontal lobes** are the parts of your brain that are responsible for thinking and are located in the brain's anterior section, on the outer surface. The major components of these lobes are the prefrontal cortex, the primary motor and premotor cortices (frontal lobes substructures located at the very front of your brain) and Broca's area (the latter three items will be talked about later).

The **prefrontal cortex** (PFC) is responsible for executive functioning (i.e., the things you choose to do; those of which you are the *executive*. The processes that occur in the prefrontal cortex include the following:

> 1. Anticipating the future – your ability to guess what your birthday present might be comes straight from your PFC.
> 2. Attention (both focusing attention and maintaining prolonged attention) – you can thank your PFC for not diverting your focus every time someone sharpens their pencil during a test.
> 3. Making plans by organizing information and setting priorities.
> 4. Decision-making – red shirt or blue shirt? Long sleeved or short sleeved?
> 5. Reflection – pondering life's mysteries.
> 6. Enthusiasm, persistence, and motivation.

Biological structure is almost always closely related to function. This holds true for many parts of the brain; the PFC is no exception. Within the maze of the brain are certain connective paths of neurons that allow signals from one part of the brain to be transmitted to another. Some of the PFC's connections, and their related functions, follow:

> 1. The left and right hemispheres are connected by a body called the **anterior commissure**. This physical connection, which bridges the right and left PFCs, allows them to work together by doing separate tasks simultaneously.
> 2. The frontal lobes span a wide range; in order for distant parts of the frontal lobes to communicate with one another, as well as with surrounding structures, some sort of connection is needed – and the **uncinate fasciculus** does just that. This connection lets the amygdala (the emotion/danger processing part of the brain) come into contact with the PFC, providing the functionality of mixing emotions and sense of danger with logic and reasoning.

> 3. The PFC is connected to the spinal cord. In short, although there are many other brain regions involved, this allows conscious movement to happen.

Beneath the cortex exists an area named the **cingulate gyrus**, which has an anterior section that is part of the frontal lobes and a posterior section that is part of the parietal lobes. The cingulate gyrus has a few jobs. First of all, it's heavily involved in attention: it prioritizes what gets attention/focus, and then relays that message to the PFC. It also takes part in actions that are driven by a goal. The cingulate gyrus contains a system that assesses which factors are the most important in the process of making a decision (and once again communicates this to the PFC where the *execution* of the decision occurs).

The cingulate has some interesting connections that are directly related to specific functionality. The anterior end of the cingulate connects the frontal lobes with the limbic system, which is a grouping of brain structures often referred to as the "feeling brain". In combination with its involvement in goal orientation, the cingulate can utilize this connection to be responsible for empathy and/or feelings of attachment. Also, the posterior cingulate gyrus bridges the frontal lobes with the parietal lobes. Because the parietal lobes are generally responsible for awareness, the cingulate provides a physical connection between thought and awareness.

At the posterior of the frontal lobes, and anterior to the **central sulcus** (a gap between the front and back of the brain), lies the motor cortex. The motor cortex is split into two parts: the anterior is the premotor cortex, and the posterior is the primary motor cortex. As you may have guessed, the motor cortex is responsible for movement. The premotor cortex plans the movements, and then sends instructions to the primary motor cortex to initiate the execution. Not only is the motor cortex responsible for the planning and initiation of movement, but it also keeps track of and maintains motion.

The motor cortex has many physical connections to various parts of the brain that specialize in certain movement-related operations. These include: the frontal lobes, the somatosensory cortex, the basal ganglia, the cerebellum, the brain stem, and the spinal cord.

One of the most frequent movements you make is talking: moving your jaw and tongue while vibrating your vocal cords produces specific sounds. While various parts of the brain contribute to this physical movement, two partnered brain components that only exist in the left hemisphere (one in the frontal lobe and one in the temporal lobe) are responsible for the execution and comprehension of speech: **Broca's area** in the frontal lobe, and **Warnicke's area** in the temporal lobe.

Broca's and Warnicke's areas are responsible for different things. Here's a breakdown:

> Broca's area:
> 1. Figuring out a way to verbally express thoughts.
> 2. Turning ideas into words.
> 3. Beginning the act of talking.
> 4. Making the act of speaking flow fluidly.
> 5. Implementing correct grammar.
>
> Warnicke's area:
> 1. Receiving ideas from others' speech by translating words into concepts.
> 2. Analyzing what others say.
> 3. Injecting words with meaning.

These two areas are intricately connected, allowing the "coming in" and "going out" parts of communication to interact smoothly.

The second of the four lobes are the **parietal lobes**. Located in the posterior of the brain, directly above the cerebellum, each parietal lobe has three main parts (of which we will only discuss the first one): the somatosensory cortex, the superior parietal lobule, and the inferior parietal lobule.

The parietal lobe, in general, is responsible for the awareness you have of your surroundings, playing a role in attention, the analysis of what's around you, and mathematical operations. The parietal lobes are connected to the thalamus, and the cortex, allowing the exchange of sensory information and information about attention and analysis. The parietal lobes output information to the frontal lobes, the cingulate gyrus, the insula (to be talked about later), and the occipital lobes.

The most important part of the parietal lobes is the somatosensory cortex (SC). This structure is located directly behind the central sulcus, and above the thalamus. The somatosensory cortex evaluates every bodily sensation except for smell, and creates a body image. Interestingly, the SC creates a "body map" of sorts by giving each body part a specific location in the structure. The SC sends information about your surroundings to the motor cortex, and disseminates the environmental information to other locations to which the information is relevant.

The third of the cortex's lobes are the **occipital lobes**, which are located in the back of the brain, posterior and inferior to the parietal lobes, but inferior to the cerebellum. The primary role of this brain section is to deal with visual processes. These are:

> 1. Receiving sensory information from the eyes.
> 2. Gathering and packaging the sensory information (perception).
> 3. Making sense of visual information (interpretation).
> 4. Allowing other parts of the brain access to the visual sensory input (dissemination).

The occipital lobes receive sensory information from the thalamus, the location where your eyes initially send their signals along an optical nerve. The processed visual data is then sent out to the somatosensory cortex, frontal lobes, temporal lobes, hippocampus, and amygdala to be used for environmental awareness, decision-making, memory, and emotion respectively.

The fourth and final set of lobes are the **temporal lobes**. In the front of the brain, inferior to the frontal cortex, the temporal lobes contain the primary auditory cortex, the secondary auditory cortex, Wernicke's area (which we discussed above), the hippocampus, the amygdala, and the fusiform gyrus. These will be discussed in detail later on.

The temporal lobes compile auditory information together, interpret that compilation, and then send it off to the rest of the brain. There are robust interconnections between the six portions of the temporal lobes, allowing the efficient passage of auditory information between these processing sections of the brain. The temporal lobes, similar to the occipital lobes, receive input from the thalamus, and also send information to the frontal lobes, the hypothalamus, and the insula (thought and reaction centers).

Subcortical Structures
There are many structures in your brain underneath the four cortical lobes, but whose functions are no less important.

The **insula** is one such structure. Located below the cortex, it is inferior to the frontal lobes, next to the temporal lobes, and anterior to the parietal lobes. The insula is involved in a wide range of functions, including:

> 1. Monitoring the state of the body, and making adjustments according to the environment so that homeostasis can be maintained.
> 2. Regulating the autonomic nervous system.
> 3. Regulating appetite and eating.
> 4. Being responsible for taste.
> 5. Being involved with intuitions, instincts, and gut feelings.
> 6. Integrating thoughts and feelings.
> 7. Injecting emotion into language.
> 8. Experiencing pain.
> 9. Feeling disgust.

The insula is located in a spot that gives it connections to a wide array of brain structures. It is connected to many major parts of the brain, including the amygdala, the frontal lobes, and the hypothalamus. Because the insula and the amygdala have a connection, being the regulator of the autonomic nervous system (reaction center), the insula can cut off an emergency signal from the amygdala to stop the body from reacting. Separately, the insula's connection to the frontal lobes and the hypothalamus (thought and homeostasis) allows it to communicate effectively about appetite.

Another important subcortical structure is the **corpus callosum**. This brain component is especially important because, being located in the middle of the brain, it connects the two hemispheres. The corpus callosum is comprised of three bundles of connecting fibers: the anterior commissure, the body of the corpus callosum, and the posterior commissure.

Due to its central location, the corpus callosum is able to be a communicator between the two halves of the brain, as well as between distant parts of the same hemisphere. The most significant effect of this is that the same components in different hemispheres can specialize in certain tasks; there is no accidental redundancy because they can communicate and coordinate.

Also lying beneath the cortical structures is the **hippocampus**. Although technically part of the temporal lobes, the hippocampus is often talked about as a separate structure due to its unique functions. The hippocampus is used in the learning of new information, as well as the creation of explicit memories – memories that we "know" we have. The hippocampus has a few important connections that allow its functions to be integrated with other parts of the brain. The heavily fortified connection between the hippocampus and the frontal lobes gives the thinking brain access to memories, meaning that when we go around and make decisions in our daily lives, we have conscious access to things we have experienced in the past. The hippocampus' connection to the thalamus allows for the direct input of sensory information into memories (due to the thalamus being the initial relay point for incoming sensory inputs), and most interestingly, the hippocampus has a strong connection to its next door neighbor, the amygdala, which creates a constant connection between memory and emotion, perhaps one of the more noticeable and relatable connections in your brain.

The **amygdala** also serves an important role. It is on the inner side of the temporal lobe, right in front of the hypothalamus. It's responsible for becoming aware of threats, and once a threat has been detected, activating the response you know as stress. Additionally, the amygdala is involved with the learning of implicit memories (things you don't have to consciously "know" to take advantage of – such as the layout of your home).

The amygdala has many connections to the frontal lobes, the cingulate gyrus, the insula, the somatosensory cortex, and other structures that carry out higher-order functions. An interesting example of how brain structure plays out into your experience with the world is that there are 10 times more connections from the amygdala to the frontal lobes than in reverse, which demonstrates structurally why emotions and fear seem to override rational thought more often than not.

Parts of the limbic system (the emotional brain) such as the nucleus accumbens (to be talked about below), the hippocampus, and the hypothalamus are connected to the amygdala. This creates fortified emotional associations to stimuli, memories, and reactions. As part of the reaction connectivity, the amygdala can communicate with the brain stem (in particular, the reticular formation) as well. This lets the amygdala's threat detection regulate things such as blood pressure and heart rate – both of which non-coincidentally rise when you are stressed.

As was mentioned above, the **nucleus accumbens**, lying inferior and posterior to the frontal lobes, and anterior to the corpus callosum, is a part of "the feeling brain" as well. This brain component is heavily responsible for the reward system and reinforcement via rewards – and consequently, drive and motivation.

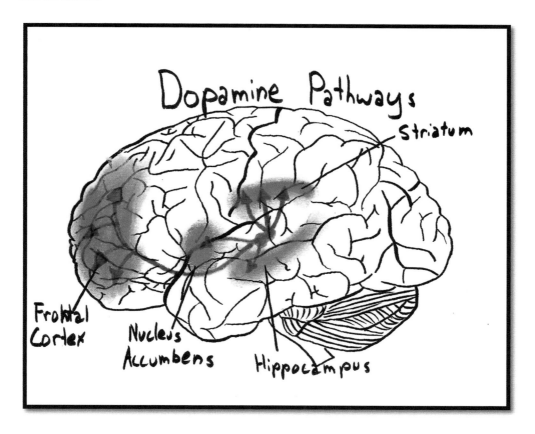

The nucleus accumbens produces and stores dopamine, a neurotransmitter whose presence we experience as "feeling good."

The nucleus accumbens is involved in the pathways of several different processes, including thinking/working memory and focus, mood and memory, and movement.

Right next to the nucleus accumbens are the **septal nuclei**. It's generally thought that this part of the brain is involved with sexual reward and rage, and has back and forth connections with the nucleus accumbens, and other parts of the limbic system. This links sexual drive and rage with dopamine rewards and emotion processing.

On both sides of the thalamus and towards bottom of the brain's anterior are the **basal ganglia**, another important subcortical structure. They are small bundles of neurons that are split into the putamen and caudate nuclei (which are referred to collectively as the striatum), the globus pallidus, and the substantia nigra. This collection of ganglia develops memory for actions such as riding a bike, which is referred to as procedural memory. The benefit this gives us in our everyday operations is that it allows us to focus on higher-order things such as thinking while still moving. If you have ever had a conversation while walking, your basal ganglia are to thank.

One of largest subcortical structures is the **cerebellum**, which resides at the posterior of the brain, inferior to the occipital lobes. The cerebellum coordinates movement and balance, which involves making movement smooth and precise, as well as combining motor, cognitive, and sensory functions.

Your brain's cerebellum is connected to many locations in the brain. Two of the most notable are the basal ganglia and the frontal lobes. By being connected to the frontal lobes, the cerebellum can help coordinate newly learned movements, and by being connected to the basal ganglia, it can fine-tune memorized motions. The cerebellum is also connected to the pons and medulla, which are part of the brain stem. These two structures are involved with involuntary movement, and the cerebellum helps them function smoothly.

When your brain receives sensory information, it all (except for smell) goes through a relay station: the **thalamus**. In the middle of the brain, directly below the cortical hemispheres and the corpus callosum, the thalamus sits and receives raw, unprocessed information from sensory organs. It then filters and focuses on pain as well as other physical sensation, and chooses which sensations to relay to both cortical and subcortical brain components.

The cerebellum, basal ganglia, and spinal cord all input information to the thalamus; most notably, the spinal cord funnels the sensory information that the thalamus distributes to the proper brain regions.

Directly below the thalamus is the **hypothalamus**. It is made of nuclei, or small segments of grey matter. The hypothalamus is a key player in emergency responses and the maintenance of homeostasis, which includes reproduction and growth. Getting input from sensory regions via the thalamus, it sends information directly to the pons and medulla, which are parts of the autonomic nervous system (involuntary actions). By doing so, the hypothalamus can regulate body functions. Additionally, the hypothalamus is connected to the pituitary gland, which release hormones for emergency response, growth, and reproduction.

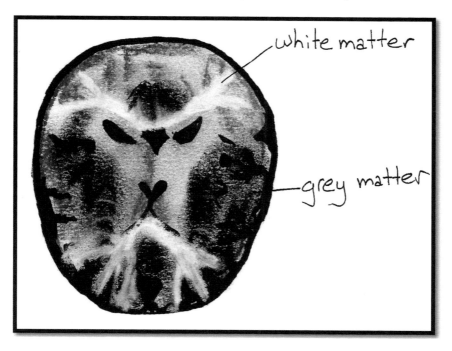

Right below the hypothalamus, the **pituitary** gland is in front and on top of the brain stem. When the hypothalamus decides there's a need for some sort of regulatory action, it signals to the pituitary gland to release hormones, directly affecting the body's actions.

The Brainstem
At the base of your brain is the brainstem, which is primarily responsible for movement and other bodily functions that need information relayed.

At the top of the brainstem are the **pons** and **medulla,** two separate brain components (the pons is above the medulla) that are directly below the thalamus. These function as a communication network between the cortex, the thalamus, and the cerebellum. They take bodily information and relay it to the thalamus, and take cortical information and relay it to the body.

Some of the pons and medulla's most important functions are the regulation of heart rate, sleep and waking, and breathing, as well as pain control. The medulla, specifically, triggers vomiting. The pons and medulla are also connected to 12 cranial nerves, each responsible for a different sense and/or movement (for example, the first cranial nerve carries information about smell, and the eighth cranial nerve carries information about sound and head orientation).

The **autonomic nervous system** (ANS) and the **reticular formation** are the other two main components of your brainstem. The ANS has two parts, the sympathetic nervous system (SNS) and the parasympathetic nervous system. (PNS) The SNS is located on either side of the spinal cord in the lumbar and thoracic regions and is made of lumps of grey matter. The PNS is in the sacral spinal cord, and is made up of knots of neurons as well. The reticular formation is a bundle of small structures that begins at the top of the spinal cord and extends downwards.

The SNS is activated for emergency responses, and the PNS is activated for homeostasis: telling the body when and how to conserve resources and energy. As a whole, the reticular formation maintains the functions necessary for life to be supported, is the direct regulator of sleep, and is thought to create the experience of being conscious. All of these structures are highly connected in the brain, as they are the implementation devices of various commands originating in all different parts of the brain.

The Cells in the Brain

We have talked about the larger-scale structures and functions of the brain. However, all of the regions are made of *something*...so what is that something, and how does it work?

The central nervous system, which includes the brain and nerves throughout your body, are made of **neurons** and glia. Neurons are the basic processors of information. There are up to 200,000 of them, and 10,000 different types! They send and receive information between themselves. Each neuron has somewhere between 5,000 and 200,000 connections.·

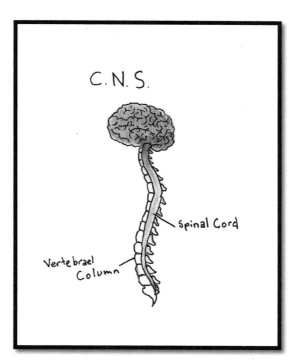

Glial cells exist in a much greater abundance than neurons do. In fact, for every neuron there are about 50 glial cells.· They create the structural framework for the brain regions, and help perform upkeep on the neurons by providing various necessary compounds and removing waste, among other things. Glial cells are more or less the housekeepers for neurons.

We are going to focus on neurons, because they're directly involved in the activities of the brain. There are four basic parts of a neuron: the cell body, the dendrites, the axons, and the axon terminal. The cell body synthesizes proteins and contains DNA, the basic encoding structure of all life. **Dendrites** receive incoming signals, and there are generally a large number of dendrites per neuron. **Axons** transmit outgoing signals, and each neuron has only one. The **axon terminal** is at the end of the axon, and contains **neurotransmitters** (the chemical substances that allow signals to be transferred from one neuron to another).

The brain is an immensely complex network, and each neuron must be able to communicate with the neurons around it. To do so, interactions between the cells occur in processes known as neurotransmission and synaptic transmission. Communication within the individual neurons themselves, in contrast, is called **conduction**.

When conduction occurs, an action potential (potential for an electrical signal to be produced) is made on the axon near the cell body. In the axon, ions (charged particles) move through ion channels – gaps in the cell membrane which are opened and closed by neurotransmitters. Normally, the charge on one side of this membrane (inside the axon) is different than the charge on the other side of this membrane (outside the axon terminal). This is called being polarized. When the ions are transferred through the membrane, the inside of the neuron becomes more positive (depolarized). As the depolarization reaches a certain threshold, an electrical signal is created – an action potential. This whole process travels down the length of an axon at up to 150 m/s.[vii]

This electrical signal, which is the "information" carried by neurons, must then be transmitted to another cell. This communication occurs as either **neurotransmission** or **synaptic transmission**. In neurotransmission, a electrical synapse is created where ions move from one neuron to another through channels called gap junctions. In synaptic transmission, chemicals are moved between neurons. This happens at locations called chemical synapses, which are a connection between an axon and a dendrite. There is a small gap between these two neuronal extensions that is filled with fluid, and is known as a synaptic cleft. The presynaptic and postsynaptic neurons (sending and receiving, respectively) compose the sides of this gap. Neurotransmitters carry ions through the synaptic cleft, continuing the electrical signal that traveled down the axon in conduction, thus allowing the neurons to communicate with one another.

In Conclusion

What has just been discussed is a broad, incomplete overview of the brain. The 200-or-so billion neurons and even larger number of supporting glial cells cannot be discussed in any small amount of space. In fact, researchers are constantly learning about new connections and functions that manifest in similar ways to the ones we have pointed out.

You will likely be relieved to hear that from this point forward, this book focuses much less on the basic biology, and more on the application and relevance to your life. Your brain, quite literally, controls your every move. It makes you happy, it makes you sad, it makes you angry, it lets you read this very text. The aim of this book is to explain how certain factors influence the structures and correlating functions discussed in this chapter within a context that, for lack of a better term, matters.

Read on, take good note, and enjoy the fascinating world of your brain and the interactions it has with your environment.

Avery Bedows

<u>Terms:</u>

Forebrain -
The top part of the brain, including the cerebral cortex, the thalamus, the hypothalamus, and other subcortical structures.

Midbrain -
The section of the brain connecting the forebrain and the midbrain.

Hindbrain -
The brain stem and its components, along with the cerebellum.

Cortex -
The outer layer of the brain, consisting of four lobes.

Frontal Lobes -
A section of the cortex that is involved with thought and motion.

Prefrontal Cortex -
The specific location of the frontal lobes that deals with thought and decision-making.

Anterior Commissure -
A cranial body that connects the right and left hemispheres.

Uncinate Fasciculus -
A brain component that helps keep the amygdala in contact with the prefrontal cortex.

Cingulate Gyrus -
A body that has parts in both the frontal and parietal lobes, prioritizing attention and creating motivation.

Central Sulcus -
A gap between the anterior and posterior of the brain.

Broca's Area -
The brain area located in the frontal lobes which is responsible for the expression of thoughts through speech.

Warnicke's Area -
A structure in the temporal lobes that processes others' speech.

Parietal Lobes -
A lobe of the cerebral cortex which deals with awareness of surroundings and mathematical operations.

Occipital Lobes -
A segment of the cerebral cortex that is responsible for processing visual stimuli.

Temporal Lobes -
A section of the cerebral cortex that is heavily involved in auditory processing, and contains parts of the brain involved with emotion.

Insula -
Located beneath the cortex, the insula is responsible for many regulatory functions.

Corpus Callosum -
A structure that provides a connection between the brain's right and left hemispheres.

Hippocampus -
Part of the limbic system, a brain structure that consolidates explicit memories.

Amygdala -
A brain component that lies directly next to the hippocampus, and is involved with fear and other emotions.

Nucleus Accumbens -
A part of the limbic system that is the brain's reward system.

Septal Nuclei -
Located next to the nucleus accumbens, this brain structure is involved with rage and sexual desire.

Basal Ganglia -
A collection of grey matter knots that is responsible for procedural memory.

Cerebellum -
At the back of the brain, directly above the brain stem, the cerebellum both makes movement more fluid and helps combine various functions.

Thalamus -
The relay center for incoming stimuli.

Hypothalamus -
Below the thalamus, this section of the brain controls homeostasis and emergency response.

Pituitary -
Connected with the hypothalamus, the pituitary releases hormones to control homeostasis, emergency responses, and growth and reproduction.

Pons and Medulla -
Large components of the brain stem, these structures help communicate between the cortex, the thalamus, and the cerebellum, as well as control some basic bodily functions.

Autonomic Nervous System -
Located along the spinal cord, the autonomic nervous system acts as the brain's communication pathway to the body.

Reticular Formation -
Part of the brainstem, the reticular formation maintains functions necessary to sustaining life.

Neurons -
The building blocks of the brain.

Dendrites -
A neuron component that receives incoming information.

Axons -
A part of a neuron that sends communications to other neurons.

Axon Terminal -
The end of an axon which is the connection point with a dendrite.

Neurotransmitters -
Chemical compounds in the brain that help neurons communicate with one-another.

Conduction -

A process by which information is transferred down an axon.

Neurotransmission -
The transmission of an electrical signal between neurons.

Synaptic Transmission -
The communication between neurons by way of chemically-supported electrical signal.

References

Bailey, R. (n.d.). Anatomy of the brain. *About.com*. Retrieved September 23, 2013, from http://biology.about.com/od/humananatomybiology/a/anatomybrain.htm

Bailey, R. (n.d.). Anatomical direction terms and body planes. *About.com*. Retrieved September 23, 2013, from http://biology.about.com/od/anatomy/a/aa072007a.htm

Davie, G. L., & Berringer, D. (2008). Base of skull tumors: types, treatment & unique rehabilitation considerations. *Texas Speech-Language Hearing Association*. Retrieved September 24, 2013, from http://www.txsha.org/_pdf/Convention/08Convention/Speaker%20Handouts/Davie,%20Gail-Base%20of%20Skull%20Tumors.pdf

Nunn, K. P., & Hanstock, T. (2008). *Who's who of the brain a guide to its inhabitants, where they live and what they do*. London: Jessica Kingsley Publishers.

Stufflebeam, R. (2008). Neurons, synapses, action potentials and neurotransmissions. *Consortium on Cognitive Science Institution*. Retrieved September 22, 2013, from http://www.mind.ilstu.edu/curriculum/neurons_intro/neurons_intro.php

Avery Bedows

[i] Nunn, K. P., & Hanstock, T. (2008). Who's who of the brain a guide to its inhabitants, where they live and what they do. London: Jessica Kingsley Publishers.

[ii] Ibid..

[iii] Bailey, R. (n.d.). Anatomical direction terms and body planes. *About.com*. Retrieved September 23, 2013, from
http://biology.about.com/od/anatomy/a/aa072007a.htm

[iv] Nunn, K. P., & Hanstock, T. (2008). Who's who of the brain a guide to its inhabitants, where they live and what they do. London: Jessica Kingsley Publishers.

[v] Stufflebeam, R. (2008). Neurons, synapses, action potentials and neurotransmissions. *Consortium on Cognitive Science Institution*. Retrieved September 22, 2013, from
http://www.mind.ilstu.edu/curriculum/neurons_intro/neurons_intro.php

[vi] Ibid.

[vii] Ibid.

The Teenage Brain: A Time of Change

From the outside, teenagers in their later years of adolescence – often called "young adults" – may appear fully grown. However, within their heads lies a different story: their brains are not entirely mature yet. The human brain does not reach maturity until, on average, 25 years of age.

By the time a child reaches the age of 5 or 6, 95% of their brain is formed, but only 80% of the brain's connections are fully developed, meaning that much growth will occur in the coming years.[viii] We can easily observe the changes in behavior that occur during adolescence, and in this chapter, we will explore the neuroscience of the changing brain and how it affects the way you act.

The Changes
Through puberty and into your early twenties, your brain is constantly developing from a child's brain to an adult brain. Right before the onset of puberty, a large number of **synapses** (connections between neurons) are formed in the brain. Although there are many connections, they are not yet efficient at transmitting information. As humans enter adolescence, a process called **synaptic pruning** occurs. In this process, important synapses are kept, while a large number are discarded, leaving more efficient information transfer routes. This ultimately leads to an increase in certain cognitive abilities.

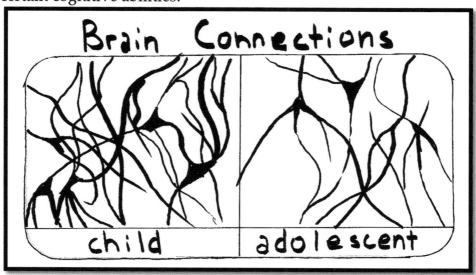

Although the brain functions better as an adult in many ways, having an excess number of synapses, like you do as a child or a young teenager, has benefits as well. Namely, an abundance of synapses in the brain leads to a remarkable ability to learn – the ability of the brain to reconfigure and create connections with stimulation. For instance, a child will find it significantly easier to learn an instrument than an adult because their brains are more frequently creating new connections in response to their environment. This comes with immense vulnerability as well. If a child endures abuse in which they do not get adequate environmental stimulation, their brains will miss the key period of building connections which will have a lifelong impact. So, if you want to learn an instrument, chess, or a sport, start now!

Young children are not the only ones who possess this remarkable ability to rearrange their brain wiring to accommodate for serious injury. Human brains, no matter their age, always have **neuroplasticity**. Neuroplasticity is a term that encompasses the various ways in which brain has the ability to change based on environmental input, including synaptic plasticity and non-synaptic plasticity. Synaptic plasticity involves synapses and the influence that use/lack of use has on them, whereas non-synaptic plasticity involves changes within the neurons themselves. Not as much is known about the role of non-synaptic plasticity in lifelong brain development, so we will primarily focus on synaptic plasticity.

Neuroplasticity usually occurs in two ways: changes in the brain during development (learning/memory), or changes in the brain that compensate for injuries.

The neuroplasticity that is a part of learning happens when existing cells grow new connections. These can be the changing of dendrite formations, the creation and destruction of synapses (synaptic plasticity), the production of new axons, or synaptic pruning, which was talked about above.

The other type of neuroplasticity is called neurogenesis. In neurogenesis, new neurons are formed. This doesn't just occur during development; it occurs all throughout your life.

While you will always have a certain amount of neuroplasticity, synaptic pruning occurs mainly during adolescence. This increased neuroplasticity explains why adolescents are usually better at learning new things. Their brains are much more malleable, meaning new connections that can store information can be formed more quickly, and in greater volume.

Where does most of this change occur? The frontal cortex. The frontal cortex undergoes tremendous change during your teenage years. Throughout the stages of brain development, the cortical lobes prune and reach maturity gradually, from the back of the brain to the front (occipital lobe first, frontal last). This is significant because it explains why certain functions in teenagers, such as visual processing and various motor actions, are equal to those of adults, while others, such as reasoning and judgment, are not.

The frontal cortex is the area of the brain that undergoes the greatest changes during adolescence. One of the two primary changes occurring here is a process known as **myelination**. Myelination is the process through which axons, which are the long parts of neurons used to establish connections, become covered with a fatty substance called **myelin** (white matter). This myelin acts as an insulator for the electrical impulses traveling through axons, preventing misfires (neurons firing when they're not supposed to). It also lets the newly myelinated axons transmit information up to 100 times faster than before.[ix]

The second large change that the frontal cortex undergoes during adolescence is synaptic pruning. This process, mentioned above, is essential to the function of the frontal cortex because it establishes greater efficiency in processing information. This, in turn, leads to improvement in the function of this region of the brain, namely reasoning and risk assessment.

The hippocampus and the amygdala are two brain structures that also undergo significant change during your teenage years. Connections between the hippocampus, which controls memory, and the frontal cortex of the brain, which deals with goals and agendas, become stronger during adolescence, allowing memory and experience (the primary functions of the hippocampus) to be integrated with the decision making process more easily. In effect, this translates into adults being better at applying previously learned information when making new decisions. For example, you probably wouldn't watch a horror movie again if you couldn't fall asleep after watching one before.

In addition to its structure, the function of the amygdala changes during the transition from child to adult as well. It appears that as your brain matures, the amygdala, a structure that deals with fear and emotion, becomes less involved in thought processes and decision making. This indicates that decisions, as you get older, become less and less influenced by emotions, and more and more rational. The increased rationality is usually viewed as beneficial, as it comes with the ability to view a bigger picture when making decisions, rather than simply being able to reference an emotional state at the exact time of a decision. And it probably is, however, there is a value to listening to "gut" feelings when making decisions. What many people view as a random intuition that has no reason, probably does have some validity. Intuition and "gut" feelings are actually your brain drawing on past experiences to subconsciously influence a decision. A youthful reliance on intuition can be useful.

The Difference Between Adolescents and Adults

The differences between adolescent and adult brains mainly reside in the frontal cortex, corpus callosum, amygdala/frontal cortex connections, hippocampus/cingulate connections, and the presence of dopamine.

Adults are almost always thought to be better at making decisions and weighing risks. But research indicates that adolescents, rather than ignoring or downgrading risks, simply weigh rewards heavier than adults when deciding whether or not to take a risk. This may be attributed to adolescents' increased sensitivity to dopamine, a neurotransmitter which appears to fire reward circuits within the brain. Researchers have begun to think that this gives humans an evolutionary advantage. Adolescents' urge to find reward forces them out of social norms and into new discoveries, eventually creating societal change.

However, teens do have a very high response to their impulses. This could be because teens do not possess fully myelinated neurons in the frontal cortex like adults do. This helps adults control their impulses much better than adolescents, which is possibly the most noticeable difference between adults and adolescents.

The **corpus callosum**, a subcortical brain structure that connects the two sides of the brain, grows during the beginning of puberty, but then stops. This structure is believed to be essential to the process of learning a new language, and its stoppage of growth during adolescent years explains why people over the age of 12 have a much more difficult time learning languages than younger children.[xi]

Another element of the adult brain that differs from the adolescent brain is the manner in which the amygdala and frontal lobes are used in facial recognition. Studies have demonstrated that when adults are shown faces, they use their frontal lobes to recognize them. But, when adolescents are shown faces, the part of the brain that appears most active is the amygdala.[xii] This points to the potential conclusion that teens seem to have more of a "gut" (emotional) reaction to recognizing people, whereas adults' reactions are more logical and based on actual analysis of information.

The connection between the hippocampus and cingulate gyrus is another component of the brain that is significantly different between adults and teenagers. The cingulate is a structure that connects the "emotional brain" (the limbic system) with the "thinking brain" (the frontal cortex). Because the connection between the hippocampus and the cingulate becomes more myelinated as adulthood is approached, memories from the hippocampus (which is part of the limbic system) become more influential on the frontal cortex. The cingulate serves the purpose of using past experiences to help determine which factors are most important in the decision-making process, and the fortification of the connection between the cingulate and the brain's memory structure gives past experiences a heavier influence on which factors are more highly valued.

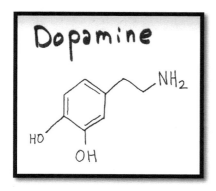

The last of the major differences between adolescent and adult brains is the presence of **dopamine**. Adolescence brings an increase in the amount of dopamine present in the frontal cortex. This assists the decision-making process by implementing memories and prior experiences when there are difficult choices to make. Dopamine, a reward chemical, helps connect choices to past experiences, which may influence a current decision.

Nature versus Nurture

One of the largest debates in the biological sciences is the ever-present question of nature versus nurture. Which characteristics are genetic? Which are influenced by the environment? To what degree do each of these factors contribute to an organism?

In this section, we'll try to tackle this question from a neurobiological perspective. There are some basic connection pathways established in the brain by your genes – no one is debating this. The question is, how does the incredibly complex neural network come to be? That question gets to the crux of "nature versus nurture." Although many parts of the brain are undoubtedly influenced by both genetic AND environmental factors, the impacts that nature and nurture have on some structures/processes is heavily debated. The structures/processes that we will address are: the function of the cerebellum, striatum, and corpus callosum, as well as the stress response, the urge to thrill seek, and neuroplasticity.

The **cerebellum**, a region of the brain that deals with motor coordination, appears to be heavily influenced by environmental input. It seems that fine motor skills such as drawing or playing an instrument may not only be genetically impacted – it appears that practice does, indeed, make perfect. The **striatum**, a part of the brain that is instrumental in maintaining a balanced presence of reward chemicals such as dopamine, is affected both by genetics and by the environment. Certain people are genetically predisposed to have a larger or smaller amount of dopamine activity, but environmental factors such as drug abuse or emotionally traumatic experiences can affect this as well. The corpus callosum, a structure involved heavily in language learning, has been shown to be primarily controlled by genetic factors, indicating that an affinity for languages comes more from genetics than environmental stimulus.

During adolescence, the presence of stress hormones is largely impacted by "nurture." Large amounts of difficult school work, increased responsibilities, and many other things trigger the production of stress hormones such as **cortisol**. When stress hormones are released, rational thought may be become overridden by the amygdala's instinctual and emotional responses. The ways in which adolescents approach problem solving when they're under stress can be affected negatively due to over-activity of the amygdala; this could be the reason you might perform worse on tests than classwork!

Because the secretion of cortisol increases the usage of the brain's emotional functions, the amygdala and the hippocampus must work extra hard. As a result, there is less dopamine present in the prefrontal cortex, and this decrease in dopamine impairs the important cognitive functions that are performed in the PFC. Here, factors that induce stress are clear examples of environmental stimuli that trigger a response by the brain.

Teenagers are well-known for being thrill seekers. It turns out that this is influenced by a combination of "nurture" and "nature." In adolescents, the neurotransmitters dopamine, epinephrine, and norepinephrine are more excitable than in adults or younger children. These neurotransmitters react strongly to the presence of stress chemicals, which as we mentioned above, are triggered by environmental stimuli.

Because these chemicals, which are involved in reward responses (dopamine is released as a reward, and epinephrine and norepinephrine are both part of thrilling fight-or-flight reaction), are so sensitive to the environment during adolescence, teenagers tend to seek thrills. A rapid dopamine increase can be a good feeling, and this is often caused by risky behaviors – which in turn cause the secretion of epinephrine and norepinephrine. Adolescents' tendency to thrill seek, as it turns out, is influenced both by environmental and genetic factors. The jittery, "pumped-up" feeling many adolescents experience while paintballing, ziplining, water skiing, or participating in other thrilling activities is the release of epinephrine and norepinephrine, and the pleasure you feel when you succeed at activities such as these is the release of dopamine.

The last part of the nature versus nurture debate which we will address revolves around neuroplasticity. Throughout adolescence, and the rest of your life, neuroplasticity accounts for the changes that your brain will undergo and is a phenomenon that is inherently curated by nurture. The changes that happen in your brain are a direct result of stimuli input. In fact, brain cells that aren't used will atrophy (become destroyed). When a large number of cells atrophy, it can be generalized that the brain "shrinks," and this phenomenon is measurable using an MRI.

One fascinating aspect of neuroplasticity is its potential for overcoming brain trauma. In a small trial study in Germany, seven patients who could not walk were placed on a treadmill, with physical support. Their legs were forced to move by the treadmill. Over time, three of the patients became able to walk by themselves, and three could walk with assistance.[xiii] This demonstrates the impact that the environment has on the brain. The neurons connected to these patients' leg function were repeatedly stimulated by the forced walking motion, and gradually adapted to the input until control was regained.

Synaptic pruning, which is also called "use-it-or-lose-it" pruning, is another clear example of how neuroplasticity is influenced by the environment. When this pruning occurs, genes activate protein synthesis that causes changes in cells. Hormonal factors are most likely involved in this process as well. However, as the name suggests, the way in which synapses are selected to be kept or discarded is based on whether or not they are actively used. Synapses that are used often will be solidified, while those that are rarely activated will be discarded. Although synaptic pruning is a genetic feature of the adolescent brain, which synapses make it through the refining process is a product of environmental input.

There is no clear-cut winner of the nature-versus-nurture influence debate. Certain functions are genetic, certain are environmental, but the vast majority seem to be a combination of the two. In adolescents, these manifest in such a way that unique phenomena occur which are seen neither in young children nor in adults.

In Conclusion
During our teenage years, our brains undergo a series of changes that give us some advantages, and some disadvantages. The inherent neuronal malleability that comes with synaptic pruning leads us, as teens, to be able to learn significantly more effectively. However, because our frontal cortices, as well as other structures and connections, are not fully developed, teenagers tend to make dangerous decisions, either because of a still-maturing risk-assessment system, or because of an increased appetite for thrilling activities.

Regardless of the positive or negative connotations of undergoing this process of change during adolescence, this is the period in our lives when we most need nutrients to help our brains develop properly and to their fullest potential. In the next chapter, we'll talk about nutrition and the brain.

Avery Bedows

Terms:

Synapse -
Synapses are the connections between neurons in the brain. They allow neurotransmitters to be transferred between cells.

Synaptic Pruning -
Synaptic pruning is a process where an abundance of synapses in the brain is narrowed down, leaving behind only the synapses that are actively used.

Neuroplasticity -
The brain's ability, throughout life, to adapt to environmental stimuli.

Myelin -
This is a fatty, white substance that covers neurons' axons, which allow electrical impulses to be transmitted between neurons. It increases transmission speed by up to 100x.

Myelination -
This is the process of coating axons with myelin.

Cerebellum -
This is a brain structure located in the back of the brain, above the brain stem. It is involved with, among other things, fine motor function.

Striatum -
This is a structure that is part of the brain's reward system.

Corpus Callosum -
This subcortical brain structure's primary function is to connect the two hemispheres of the brain, but does many other things, including bearing responsibility for learning languages.

Cortisol -
Cortisol is known as a "stress" chemical. It is a hormone that is secreted by the brain in situations of stress, and induces responses such as the fight-or-flight instinct.

References

Barford, E. (2012, October 5). Nature and nurture teased apart in brain's reward centre. *Imperial College London*. Retrieved September 29, 2013, from http://www3.imperial.ac.uk/newsandeventspggrp/imperialcollege/newssummary/news_5-12-2012-11-43-18

Chamberlain, Linda B. (n.d.). The amazing adolescent brain: what every educator, youth serving professional, and healthcare provider needs to know. *Multiplying Connections*. Retrieved September 29, 2013 from http://www.multiplyingconnections.org/sites/default/files/Teen%20Provider%20article%20(2)_0.pdf [PDF form]

Dobbs, D. (2011, October). Teenage Brains. *National Geographic* . Retrieved September 26, 2013, from http://ngm.nationalgeographic.com/2011/10/teenage-brains/dobbs-text

Hammond, K. (2002, August 28). Neuroplasticity. *HOPES*. Retrieved October 31, 2013, from http://www.stanford.edu/group/hopes/cgi-bin/wordpress/2010/06/neuroplasticity/

Hoiland, E. (n.d.). Neuroscience for kids - brain plasticity. *Neuroscience for Kids*. Retrieved October 31, 2013, from http://faculty.washington.edu/chudler/plast.html

Kays, J. L., Hurley, R. A., & Taber, K. H. (2012). The Dynamic Brain: Neuroplasticity and Mental Health. *The Journal of Neuropsychiatry and Clinical Neurosciences*, 24(2), 118-124. Retrieved October 31, 2013, from http://neuro.psychiatryonline.org/article.aspx?articleID=1213973#Conclusions

National Institute of Mental Health. (2011). The teen brain: still under construction. Retrieved September 26, 2013 from http://www.nimh.nih.gov/health/publications/the-teen-brain-still-under-construction/index.shtml

Roaten, G. K., & Roaten, D. J. (2012). Adolescent brain development: Current research and the impact on secondary school counseling programs. *Journal of School Counseling. 10 (8)*.

Ruder, D. B. (2008, October). The Teen Brain. *Harvard Magazine*. Retrieved September 26, 2013, from http://harvardmagazine.com/2008/09/the-teen-brain.html

Spinks, S. (2000, March 9). Adolescent brains are works in progress. *PBS*. Retrieved September 26, 2013, from http://www.pbs.org/wgbh/pages/frontline/shows/teenbrain/work/adolescent.html

Weinberger DR, Elvevag B, Giedd JN. *The adolescent brain*. Washington, DC: National Campaign to Prevent Teen Pregnancy; 2005.

[viii] Ruder, D. B. (2008, October). The Teen Brain. *Harvard Magazine*. Retrieved September 26, 2013, from http://harvardmagazine.com/2008/09/the-teen-brain.html

[ix] Dobbs, D. (2011, October). Teenage Brains. *National Geographic* . Retrieved September 26, 2013, from http://ngm.nationalgeographic.com/2011/10/teenage-brains/dobbs-text

[x] University of Leeds. (2008, March 6). Go with your gut -- intuition is more than just a hunch, says new research. ScienceDaily. Retrieved August 5, 2014 from www.sciencedaily.com/releases/2008/03/080305144210.htm

[xi] Spinks, S. (2000, March 9). Adolescent brains are works in progress. *PBS*. Retrieved September 26, 2013, from http://www.pbs.org/wgbh/pages/frontline/shows/teenbrain/work/adolescent.html

[xii] Weinberger DR, Elvevag B, Giedd JN. *The adolescent brain*. Washington, DC: National Campaign to Prevent Teen Pregnancy; 2005.

[xiii] Hammond, K. (2002, August 28). Neuroplasticity. *HOPES*. Retrieved October 31, 2013, from http://www.stanford.edu/group/hopes/cgi-bin/wordpress/2010/06/neuroplasticity/

The Brain and Nutrition: Why Eating Does More Than Fill Your Stomach

It's time to indulge, so you drink a can of sugary soda. Soon after, you have an intense rush of energy. And then...the crash. We all know what this is, because we've all experienced it – it's a sugar rush. But, what is a sugar rush? Is it real, or is it just a figment of our imaginations? This is a question of nutrition, and how it affects your brain. In this chapter, we will explore not only the answer to this question, but the many ways in which your brain is heavily influenced by what you consume.

The General Consensus: Good vs. Bad

The terms "proteins," "carbohydrates," "fats," "micronutrients," and others are often thrown around when discussing healthy foods for your brain. The problem is that there are many misconceptions floating about. Between fad diets, various vitamins, and the debate on sugar, how do you know which types of nutrition are *actually* good for you? And, what do they do?

To begin, let's discuss what the general consensus does and does not consider nutritional. There are four main categories of nutritional foods: proteins, complex carbohydrates, healthy fats (polyunsaturated), and micronutrients.

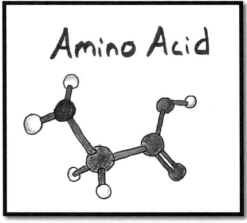

Proteins are chains of amino acids generally found in meats, chicken, fish, and certain vegetables, grains, and nuts. **Carbohydrates** are chains made exclusively of carbon, hydrogen, and oxygen. It is from these chains that our bodies create glucose, which is used to fuel our cells and keep them working properly. **Complex carbohydrates** are found in, among others, grains, fruits, legumes (different types of beans), and vegetables. Healthy fats, called **polyunsaturated fats**, are found in many different types of foods, most notably fish, other meats, vegetable oils, and dairy products. These fats, in essence, are more molecularly compact than their counterparts, **saturated fats** (which will be discussed later on).

Finally, **micronutrients** are nutrients that help orchestrate physiological functions when consumed in small quantities. They constitute a wide array of vitamins and other substances, such as vitamin D, E, F, and zinc.

Everything mentioned above is generally deemed to be of importance in a diet. However, there are plenty of foods – eaten all-too-often – that don't do your brain any favors. Most significant are refined sugars/carbohydrates, found in soda, candy, ice cream, and other sweet and sugary foods, and saturated fats. These are found in the majority of processed foods (take popcorn butter for example, or processed meats).

Proteins

Proteins are one of the four classes of nutrients discussed above that are important to have in your diet. Not only do they help build muscle and facilitate many other molecular processes in your body, but they are essential to the function of your brain for two reasons. First, proteins are required to grow new neural pathways. The brain's network of neurons requires the presence of protein in order to grow. Without protein in your diet, your brain simply wouldn't have the material to get larger and develop more connections. Additionally, proteins contain the building blocks for neurotransmitters: amino acids. This means that in order for your brain to be able to produce the proper amount of neurotransmitters, you must be getting enough protein in your diet.

It has been demonstrated in various studies that consuming protein can improve learning skills. A chronic protein deficiency would most likely lead to impaired learning abilities as a result of fewer neuronal connections being established than usual, as well as not having enough neurotransmitters present. Neurotransmitters affect emotions, so not only will learning potentially be impacted, but emotional functioning may as well. Many teenagers entirely skip out on protein at breakfast and lunch, impairing their brain function for the rest of their day.

Proteins are just one component of a diet that's good for your brain. Another major class of nutrient is the carbohydrate.

Carbohydrates
To discuss carbohydrates, we must first take a look at how brain cells get their energy:

> The energy that the brain uses to support its function is traced right back to the carbohydrates that you consume. When you eat something containing carbohydrates – say, a piece of bread – your body begins to break down the carbohydrates from your food into **glucose**, a molecular compound that is used to power your body's cells. Glucose then enters the body's bloodstream, and in turn is transported to neurons. There's a problem, however: neurons, unlike many other types of cells, are unable to store glucose. This means that there must be a constant supply of glucose in the bloodstream in order for the brain to function properly.

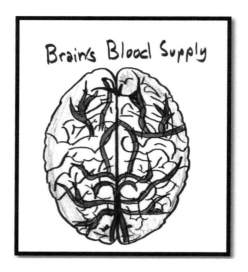

Bearing this in mind, let's look at the difference between healthy complex carbohydrates and unhealthy **refined/simple carbohydrates** (simple sugars). The word "complex" in complex carbohydrate means, quite literally, that the chain of carbohydrates is long and intricate. Conversely, refined/simple means that the chain of carbohydrates is short and has a simple molecular structure.

First, let's examine simple sugars/carbohydrates. Fructose and sucrose are the two simple sugars that we consume the most of. While fructose is a monosaccharide (a single sugar molecule), sucrose is a disaccharide (two sugar molecules).

Compared to longer chains of sugars, these two simple sugars enter the bloodstream very quickly. Complex carbohydrates are long chains of these molecules that take the body much longer to break down. As a result, when you eat complex carbohydrates, your blood's glucose level does not spike immediately after consumption. Simple carbohydrates are usually processed from complex carbohydrates. The processing strips them of vitamins and fibers (substances which resist digestive enzymes). Sugars from fruit are absorbed into the bloodstream slower than sugary fruit juice because of the fibers in the fruit.

The level of glucose in your bloodstream at any time is the amount of fuel your body and brain has at that given time. It is very important for the amount in your bloodstream to be regulated – and your body does just that.

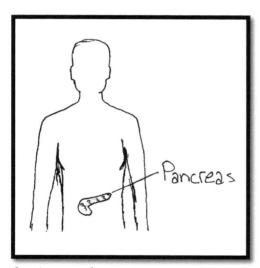

The amount of glucose transported to the brain is monitored by the pancreas. If your blood glucose level is too high, your pancreas will secrete a compound called **insulin** which reduces the glucose to a reasonable level by causing the body's cells to store it as fat. When there is a large spike in glucose in the bloodstream, the pancreas reacts and the glucose is quickly removed from the bloodstream and converted into fat. The neurons in your brain need energy on a moment-by-moment basis, so when the amount of glucose in your bloodstream drops dramatically, your brain becomes starved for energy. This is called hypoglycemia.

And now, knowing this, we can finally discuss sugar rushes. The majority of studies indicate that there is no increase in activity when children or teenagers consume large amounts of sugar. However, it has been demonstrated that after consuming glucose, certain cognitive functions such as word recall and information processing speed improve in children.[xiv]

Here's what happens biologically when you drink a large soda, or eat a lot of candy: you are consuming a large amount of *refined* sugars, and these sugars are very quickly broken down into glucose by your body. With this abundance of glucose in the bloodstream, the brain has ample energy, which potentially explains why cognitive functions are briefly boosted. The issue begins when the pancreas begins to secrete insulin, efficiently removing the glucose in your bloodstream. While the rest of the body's cells have a store of glucose, the neurons in your brain do not! With no energy reserve to pull from, they experience an energy crisis. This is what is felt as a decline in focus and cognition. In fact, when glucose levels get too low, you can faint, and in severe cases with complicating factors, enter into a coma or potentially die. This usually happens because of diabetes, which affects how your pancreas (and consequently insulin) functions. One interesting note: it turns out that if you eat protein 30 minutes before sugar, the protein helps prevent your insulin from spiking, meaning that you won't suffer from the cognitive decline shortly afterwards.

Two more interesting bits about carbohydrates:
1. Sugars cause the nucleus accumbens to release dopamine (a reward chemical). Some studies have indicated that eating excessive amounts of sugar might cause tolerance to build up in dopamine receptors, which means it is more difficult to get the sense of pleasure that dopamine usually creates.
2. Glucose reduces the hunger hormone ghrelin, and stimulates the hormone leptin, which suppresses appetite. Fructose, however, does not interact with ghrelin, and inhibits your brain from producing as much leptin as is necessary, which leads to overeating. Although fruits are high in fructose, they have fibers which help prevent it from being absorbed into the bloodstream so quickly, meaning that fruits likely don't lead to overeating in the same way that refined fructose will.

Carbohydrates, when consumed excessively, get converted into fat. Dietary fats, though, are an essential part of your diet.

Fats
Both healthy and unhealthy fats are abundant in most people's diets. The difference between these types of fats is in their molecular arrangement: unhealthy saturated fats contain no double-bonded hydrogen atoms (meaning the molecule is less

compact), whereas healthy unsaturated fats do (the molecule is more compact).

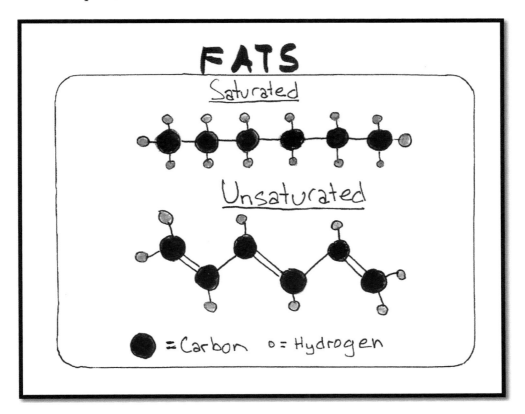

Fats play a few important roles in your brain. DHA, a type of omega-3 fatty acid, has three main functions in the brain: making cell membranes more fluid, being involved with sending signals along cell membranes, and increasing the amount of neurotransmitters in the brain.

Omega-3 fatty acids are a type of fatty acid often found in fish oil. There isn't much conclusive research on them. However, there are a series of studies that show omega-3s consumed over a long period of time have positive impacts on the brain. One study of students in Durham, United Kingdom indicated that taking omega-3 supplements regularly over an extended period of time lessened deficits in reading and spelling levels.[xv] Additionally, in a study in Australia and Indonesia, DHA and other micronutrients were given to a group of children. Compared to a control group, the children given the supplements scored higher on tests of verbal intelligence and learning, as well as memory.[xvi] One other study of rodents indicated that the combination of DHA with regimented exercise can increase spatial learning ability[xvi], a skill that has many practical applications. Although there are many aspects of these fatty acids that still must be investigated, there is significant evidence that they have considerably positive effects on the human brain.

Interestingly, some studies have indicated that adolescents who don't consume enough omega-3 fatty acids have an increased chance of developing Attention Deficit Disorder (ADD), dyslexia, dementia, depression, bipolar disorder, or schizophrenia. Basically, fish oil is a good supplement to take.

Now that we've talked about proteins, carbohydrates, and fats, it's time to discuss vitamins and micronutrients.

Vitamins/Micronutrients
Micronutrients constitute a wide array of vitamins and other substances, such as zinc and Vitamin D.

Some micronutrients have been shown to affect brain function. Zinc, an atomic element, allows the neurotransmitter serotonin to function properly. When you don't consume enough zinc, the amounts of the neurotransmitters dopamine and glutamate become unbalanced, which can lead to irritability and moodiness. It is thought that lack of zinc is one of the reasons teenagers can feel so up and down mood-wise.

Vitamin D, acquired by sun exposure, although also through diet, may be important for the brain. A study at the University of Cambridge showed that subjects who had severe vitamin D deficiencies were twice as likely to be cognitively impaired than subjects who had ample vitamin D.[xviii] Another study at the University of Manchester determined that in men across Europe, people with lower vitamin D levels tended to be slower at processing information.[xix] Although the scientific community is still unsure whether increasing vitamin D levels can reduce this correlation, it certainly seems that having enough vitamin D – either from spending time in the sun, or taking vitamin supplements – is not a bad idea.

Another class of micronutrients that have an effect on the brain are antioxidants. These substances help prevent neuronal damage from occurring due to an unavoidable natural process known as "oxidative stress," which happens as a result of certain chemical properties of oxygen. The presence of excessive oxidative stress can decrease certain types of brain function, so frequent consumption of foods containing antioxidants, such as most berries, can help prevent this damage from occurring, and improve cognitive function on a day-to-day basis.

Antioxidants have been the subject of a number of studies in recent years. Many of these studies have revealed prospective long-term positive effects of regularly consuming foods containing antioxidants. One study using animal models of Alzheimer's Disease has demonstrated that consuming antioxidants actually lessens the memory deficits created by this disease.[ix] In fact, antioxidants may explain a curious cultural phenomenon: people in India seem to have a significantly lower incidence of Alzheimer's.

> Indian food is often flavored with a spice called turmeric – which contains a chemical compound known as curcumin. Curcumin is unique in that it has a large quantity of antioxidants. There is a strong possibility that because Indians tend to consume such a large quantity of curcumin, via turmeric (Indians consume nearly 94% of the turmeric produced worldwide[xxi]), their brains suffer less oxidative stress, meaning they have a lower likelihood of developing Alzheimer's.

There are a whole host of other types of micronutrients that are good for your brain that are too numerous to mention. Needless to say though, having the proper amount of vitamins in your daily diet certainly maintains cognitive function.

One final note: a fascinating series of studies have demonstrated that in adolescents who lack certain micronutrients in their diets, taking vitamin supplements can actually boost their IQs (Intelligence Quotients).[xxii]

In Conclusion
Nutrition, both good and bad, has many effects on the developing brain. Having a diet that includes proper amounts of proteins, fats, carbohydrates, and micronutrients is a great way to set yourself up

to feel energetic and maximize your ability to learn, as well as keep your brain functioning well far into the future. When bad foods such as simple sugars and saturated fats are excessive in your diet, however, your brain won't function nearly as well, nor will it down the road.

It is essential to find a dietary balance that leaves you feeling good and feeling focused, not only so that you can perform well in your studies, but so you can maintain your mental health in the future. What you consume, however, is only one of the factors that contributes to how your adolescent brain develops. In the next chapter, we will discuss the effects of screen time on the workings of your brain.

Terms:

Proteins -
Proteins are chains of amino acids that serve many functions in the body. Protein is essential to the upkeep of brain function, as well as for the maturation of neural pathways.

Fats (Polyunsaturated and Saturated) -
Fats are a class of organic substances. There are two main types: polyunsaturated and saturated. Polyunsaturated fats are fats that contain more than one Carbon-Carbon double bond, and saturated fats are fats that do not.

Carbohydrates (Complex and Refined/Simple) -
Carbohydrates are molecules made of Carbon, Oxygen, and Hydrogen. The difference between complex and refined/simple carbohydrates is the size of the molecule, which impacts the rate at which it can be broken down by your body.

Insulin -
Insulin is a compound secreted by the pancreas which takes glucose out of the bloodstream, and stores it as fat in the body's cells.

Glucose -
Glucose is your body's source of energy. When you consume carbohydrates, your stomach will convert them into glucose, which is then distributed to cells all throughout your body to give them the energy necessary to function.

Micronutrients -
Micronutrients are a series of substances required by the human body to function properly. Items that fall under this class are vitamins (organic compounds needed by the body) and microminerals/trace elements, which are just small amounts of periodic elements.

Avery Bedows

References

Erickson, R. (2011, January 3). How nutrition affects the brain of adolescents. *Livestrong.com*. Retrieved September 8, 2013, from http://www.livestrong.com/article/364298-how-nutrition-affects-the-brain-of-adolescents/

France, B. (2004). Effects of diet on behaviour and cognition in children. *British Journal of Nutrition*, 92(Supplement S2), S227-S232. Retrieved November 3, 2013, from http://hundsundskolerestaurant.no/wordpress/wp-content/uploads/2010/11/Bellisle-sugar-and-cognition-in-children-2004.pdf

Gomez-Pinilla, F. (2008). Brain foods: the effects of nutrients on brain function. *Nature Reviews Neuroscience*, 9(7), 568-578. Retrieved September 8, 2013, from http://dx.doi.org/10.1038/nrn2421

Greenberg, M. A. (2013, February 5). Why our brains love sugar – and why our bodies don't. *Psychology Today*. Retrieved November 3, 2013, from http://www.psychologytoday.com/blog/the-mindful-self-express/201302/why-our-brains-love-sugar-and-why-our-bodies-dont

Health effects of fats: fats and the brain. (n.d.). *Fats of Life*. Retrieved November 3, 2013, from http://www.fatsoflife.com/health-effects-of-fats-fats-and-the-brain/

Health effects of fat: mental health. (n.d.). *Fats of Life*. Retrieved November 3, 2013, from http://www.fatsoflife.com/health-effects-of-fats-mental-health/

Hedaya, R. J. (2010, June 3). The teenager's brain. *Psychology Today*. Retrieved September 8, 2013, from www.psychologytoday.com/blog/health-matters/201006/the-teenagers-brain

Hendrickson, K. (2010, November 18). Fructose vs. sucrose. *Livestrong.com*. Retrieved November 3, 2013, from http://www.livestrong.com/article/311336-fructose-vs-sucrose/

Lawson, W. (2003, January 3). Brain power: why proteins are smart. *Psychology Today*. Retrieved November 3, 2013, from http://www.psychologytoday.com/articles/200301/brain-power-why-proteins-are-smart

Mercola, J. M. (2011, February 28). Fructose affects your brain very differently than glucose. *Mercola.com*. Retrieved November 3, 2013, from http://articles.mercola.com/sites/articles/archive/2011/02

/28/new-study-confirms-fructose-affects-your-brain-very-differently-than-glucose.aspx

Murphy, J. M. (2007). Breakfast and learning: an updated review. *Current Nutrition & Food Science, 3*(1), 3-36.

National Multi-Commodity Exchange of India Limited. (n.d.). Report on Tumeric. Retrieved September 10, 2013, from http://www.nmce.com/files/study/turmeric.pdf [PDF Form]

Neuroscience for Kids - Nutrition and the Brain. (n.d.). *UW Faculty Web Server*. Retrieved September 8, 2013, from http://faculty.washington.edu/chudler/nutr.html

Noortlaan, O. v. (2007, October 10). Nutrition improves learning and memory in schoolchildren . *Medical News Today*. Retrieved October 28, 2013, from http://www.medicalnewstoday.com/releases/85094.php

Northstone, K., Joinson, C., Emmett, P., Ness, A., & Paus, T. (2012). Are dietary patterns in childhood associated with IQ at 8 years of age? A population-based cohort study. *Journal of epidemiology and community health, 66*(7), 624-628.

Ramsey, D. (2012, April 17). Meat is brain food. *The New York Times*. Retrieved September 8, 2013, from http://www.nytimes.com/roomfordebate/2012/04/17/is-veganism-good-for-everyone/meat-is-brain-food

Saturated vs. unsaturated fatty acids. (n.d.). *FITDAY*. Retrieved November 3, 2013, from http://www.fitday.com/fitness-articles/nutrition/fats/saturated-vs-unsaturated-fatty-acids.html#b

The human brain - carbohydrates. (n.d.). *The Franklin Institute Online*. Retrieved September 11, 2013, from http://www.fi.edu/learn/brain/carbs.html

Welland, D. (2009, November). Does vitamin D improve brain function?. *Scientific American*. Retrieved November 4, 2013, from http://www.scientificamerican.com/article.cfm?id=does-d-make-a-difference

Avery Bedows

[xiv] France, B. (2004). Effects of diet on behaviour and cognition in children. *British Journal of Nutrition*, *92*(Supplement S2), S227-S232. Retrieved November 3, 2013, from http://hundsundskolerestaurant.no/wordpress/wp-content/uploads/2010/11/Bellisle-sugar-and-cognition-in-children-2004.pdf

[xv] Gomez-Pinilla, F. (2008). Brain foods: the effects of nutrients on brain function. *Nature Reviews Neuroscience*, *9*(7), 568-578. Retrieved September 8, 2013, from http://dx.doi.org/10.1038/nrn2421

[xvi] Ibid.

[xvii] Ibid.

[xviii] Welland, D. (2009, November). Does vitamin D improve brain function?. *Scientific American*. Retrieved November 4, 2013, from http://www.scientificamerican.com/article.cfm?id=does-d-make-a-difference

[xix] Ibid.

[xx] Gomez-Pinilla, F. (2008). Brain foods: the effects of nutrients on brain function. *Nature Reviews Neuroscience*, *9*(7), 568-578. Retrieved September 8, 2013, from http://dx.doi.org/10.1038/nrn2421

[xxi] National Multi-Commodity Exchange of India Limited. (n.d.). Report on Tumeric. Retrieved September 10, 2013, from http://www.nmce.com/files/study/turmeric.pdf [PDF Form]

[xxii] France, B. (2004). Effects of diet on behaviour and cognition in children. *British Journal of Nutrition*, *92*(Supplement S2), S227-S232. Retrieved November 3, 2013, from http://hundsundskolerestaurant.no/wordpress/wp-content/uploads/2010/11/Bellisle-sugar-and-cognition-in-children-2004.pdf

Do Computers Rot Your Brain?

Adolescents are spending more and more time online. Homework is now accessible, and often turned in, via the internet. Social networks are becoming a primary means of communication. There are a virtually unlimited number of games available online. In many people's eyes, this proliferation of the internet is considered a good thing, a hallmark of advancement in efficiency and communication. But people question how good screen time is for an adolescent brain, prompting many, many studies and some potentially premature arguments that it "rots your brain!" Is there a conclusive answer?

Screen Time

Although adolescents tend to love "**screen time**," a term given to spending time using electronics that have a screen, using gadgets too often has generally been given bad press by the media. Violent video games, in specific, have become notorious for supposedly inducing violent behavior. In this chapter we will examine the current research in the scientific community about what spending time using electronics does to the adolescent brain.

The Internet

Internet usage has exploded in recent years. Having a vast expanse of knowledge a click away is an incredible benefit. However, there is a growing worry that when adolescents use the internet excessively, issues arise.

One such issue is that in very heavy internet gamers, there have been signs of **brain atrophy** (decay). Studies have shown that neurons in adolescents' brains seem to break down at a rate that is a function of the time spent gaming. In one study, MRI's of internet addicts (an actual addiction that must be taken seriously) showed indications of atrophy in various areas of the brain, most importantly the frontal lobe which controls executive functions such as planning, organizing and impulse control. This could explain parents' claims that internet overuse makes their children behave erratically.[xxiii]

Social networks are another realm of the internet that have been found to negatively impact teenagers. First of all, through cyberbullying. Quite simply, it is much easier to be cruel to a peer from behind a screen than in real life, increasing the number of teens who feel comfortable verbally attacking schoolmates. The internet is an unregulated space that proves very difficult to avoid. This leaves adolescents with no way to fight back or escape their abuse, leading to an alarming number of suicides and episodes of depression that have stemmed from cyberbullying.

Additionally, the amount of time spent with Facebook open on the computer seems to be inversely correlated to reading retention.[xxiv] This could potentially hold implications for a connection between Facebook usage and school performance, and is worrisome given the continuously increasing adolescent membership on Facebook. However, for kids who are shy and struggle with social interaction, Facebook, and networks like it, could essentially work as social "training wheels," allowing them to interact through messages, practicing for real life.

This research is problematic because most of these studies have only been performed on adolescents who have internet addictions. 95% of adolescents do not have problematic internet use. You'll be glad to hear, there is not yet research indicating any negative effect from moderate internet usage.

Cell Phones
One of the most common access points to the internet nowadays are mobile phones. These devices, however, come with their own side effects when used excessively by adolescents. Although cellphones have undoubtedly contributed positively to our society, their continuing increase in the teenage market has yielded some fascinating new observations.

To begin, here's a question: what fraction of adolescents sleep with their phones next to them?

Answer: 4/5.[xxv]

Four in every five teenagers, when they go to sleep, place their cellphones next to them. Check it before bed, check it in the morning. This is demonstrative of the enormous attachment adolescents have to their phones, and this attachment seems to create some very interesting outcomes. When a teen actively participates in a conversation over text messaging, there is a major increase in activity in the brain's pleasure centers – the very same ones that are activated by heroin. Shocking, right? (However, don't be too terrified, chocolate activates the pleasure center as well!) Cellphones have similar addictive qualities to addictive drugs.

Cellphones also may influence the way in which children learn. Most, if not all, smartphones have a built in dictionary that is used for predictive or corrective texting. A phone will, quite often, finish the words for you before you even finish typing them. And in some cases, when a word is spelled wrong, the phone will automatically correct it.

Bearing this in mind, a team of scientists recently conducted an inquiry to determine the impact of frequent cell phone usage on the minds of 11-14 year olds by administering IQ tests and then analyzing the results. They discovered something fascinating: frequent phone users completed the test faster, but with less accuracy.[xxvi] A postulated cause of this result is predictive texting. Its presence allows children, when typing, to make a greater number of mistakes but still get the intended product. There is concern that this effect demonstrates how cellphone use by early adolescents can affect the ways in which they learn, making them more prone to inaccuracy because of constantly being corrected automatically.

Although cellphones seem to have some tangible effects on teenagers' minds, there is no type of electronic entertainment device that is more controversial than video game consoles.

Gaming: The Truth

Video games are everywhere in our society. Large game releases are advertised on television just as much as TV show premiers and upcoming films. The gaming industry has been growing massively as graphics become more and more realistic, and the games become increasingly immersive, fully engaging the gamer in their virtual environment.

Video games, however, have been the subject of innumerable studies investigating claims that, among other things, they lead children to be violent. First, however, let's examine video game addiction. It has become apparent that adolescents who are frequent gamers have larger ventral striata than adolescents who do not spend as much time engaging in these pastimes.

As part of the brain's reward center, the striata releases chemicals that make you feel good, and consequently affects desire to engage in certain actions. Irregularities in this structure can express themselves in the form of schizophrenia, Obsessive Compulsive Disorder (OCD), and addiction. In fact, the striata of extremely frequent teen gamers resembles the striata of drug addicts.[xxvii]

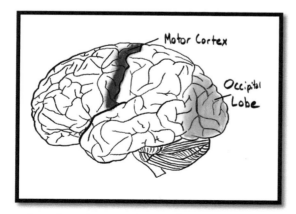

Perhaps a more pressing issue with video games is the effects that they have been shown to have on the frontal lobes. It has been observed that when you play a video game, the parts of your brain that are activated are the ones that process vision and integrate motion (occipital lobes, motor cortex, and others). A study[xxviii] has found that video games stimulate exclusively the vision and movement centers of the brain, whereas performing mathematics (a task that requires complex thought) can activate the right and left hemispheres of the frontal lobes, locations of the brain that are associated with learning, memory, emotion, and behavior. It has been contended that video games cause underdeveloped frontal lobes in adolescents who frequently play them, and thus it would follow that deficiencies in the aforementioned areas of functioning (learning, memory, etc.) would be present.

However, how exactly do video games affect the frontal lobes? That is an important question. The answer lies in the transitional characteristics of adolescent brains. As was discussed in Chapter 2, adolescent brains go through a process called "pruning." In this process, neuronal connections that are unnecessary are eliminated, and the ones that are significant to the function of the brain are kept and fortified. The key here is that this process is influenced by stimulation outside of the body.

Video games quickly overstimulate the brain, so signals don't pass through the frontal cortex (the decision center) – reactions to what's going on in the video games become habitual, and become processed explicitly by visual and motor brain structures. If this process is repeated day in and day out, as it is with chronic gamers, pruning occurs that solidifies neuronal connections which bypass the frontal cortex when responding to stimuli, meaning that adolescents' executive functions such as thinking, logic, and rationality are not utilized when responding to significant stimuli. The brain, in effect, is taught not to *think*, but to *do*. This may lead to more brash, less reasonable responses to events.

This could very well explain why frequent gamers may to respond to violent, stressful stimuli in a way that does not utilize their sense of rationality, but purely their instincts. Of course, this may be a correlation, not a causation, and those who aren't particularly rational are simply attracted to video games. It has been suggested that this could explain an increase in youth violence, but as of yet, there is little information supporting this assertion.

Video games are an interactive form of screen time; your brain has to respond. Television, however, is all of the screen-time, but none of the interaction. Let's take a look at what watching TV does to your brain.

Television

There's not much debate: watching excessive TV is NOT good for you. In fact, multiple studies have been done which shed light on what exactly watching too much TV as a child can do.

Watching television has been associated with depression. One study demonstrated that for each additional hour of television watched per day during early childhood, the chances of developing depression as an adolescent increase a massive 8%.[xix] Another study showed that people who are unhappy tend to watch significantly more television per week than people who are: 25 hours for people who are unhappy versus 19 hours for those who are.[xx]

Watching excessive amounts of TV seems to be especially bad for young children. In a study of 1,000 toddlers, over the course of maturing, the children who watched more TV were more likely to have difficulty performing mathematics, and engage in misbehavior.[xxi] This is likely because instead of doing stimulating activities that allow the toddlers to form brain connections, they were sitting and doing nothing.

Another study showed that young kids who spend more than 2 hours per day watching television are twice as likely to have attention problems as young adults![xxii]

3-5 violent acts are depicted in an average hour of prime-time television and 20-25 violent acts are depicted in an average hour of children's television. Sensibly this leads to a worry that watching too much TV causes an increase in violence. It turns out there may be truth in this. A study of 707 children that was conducted over the course of 17 years yielded the following results: of 360 male 14 year old subjects, of those who watched less than 1 hour of television a day, 8.9% committed a violent act towards another person, of those who watched between 1-3 hours of TV per day, 32.5% committed a violent act towards another person. And the children who watched more than 3 hours per day? 48.2% of them committed a violent act towards another person.[xxiii] These numbers are only just considered statistically significant, when controlled for the many other factors that induce violent behavior, such as: IQ, socioeconomic status, education, and childhood neglect. However, it is clear that for males, more television equals more violence.

Further research is needed to figure out why it is, exactly, that watching TV has such harmful side effects.

In Conclusion

Electronic devices are pervasive in the worlds of teenagers across the globe. Many young adults spend hours a day interacting with these devices, and it appears that these interactions may have some interesting, if not negative, impacts on the teenagers. However, there is a dearth of conclusive research at the moment, and a tendency for media to take disputable evidence of negative effects from devices and embellish it, until it is no longer representative of any truth. Hopefully this will change in the future. And for the time being, as a general rule, use "everything in moderation" and your brain, most definitely, will not rot away!

One element that most electronic devices share is a backlit screen. As you will see in the next chapter, however, these screens can have an impact on one your body during one of its most essential functions: sleeping.

Avery Bedows

Terms:

Screen Time -
The term "screen time" is used to refer to the use of electronic devices for purposes of entertainment, social connectivity, or education. These include, but are not limited to, computers, cell phones, televisions, and tablets.

Brain Atrophy -
The word "atrophy" means decay. In this case, the term atrophy, or brain atrophy, refers to the decay/breaking-down of neurons in the brain.

References

Bongiorno, P. (2011, October 6). Your unhappy brain on television. *Psychology Today*. Retrieved October 28, 2013, from http://www.psychologytoday.com/blog/inner-source/201110/your-unhappy-brain-television

Connor, S. (2014, May 20). Moderate internet use unlikely to harm teenagers' brains, study finds. The Independent. Retrieved August 8, 2014, from http://www.independent.co.uk/life-style/health-and-families/health-news/moderate-internet-use-unlikely-to-harm-teenagers-brains-study-finds-9404858.html

Cooper, S. (2010, October 10). 6 shocking ways TV rewires your brain. *Cracked*. Retrieved October 28, 2013, from http://www.cracked.com/article_18856_6-shocking-ways-tv-rewires-your-brain.html

Dunckley, V. L. (2014, February 27). Gray matters: too much screen time damages the brain. Psychology Today. Retrieved August 8, 2014, from http://www.psychologytoday.com/blog/mental-wealth/201402/gray-matters-too-much-screen-time-damages-the-brain

Grant, C. (2009, August 11). How predictive texting takes its toll on a child's brain. *Mail Online*. Retrieved September 11, 2013, from http://www.dailymail.co.uk/sciencetech/article-1205578/How-predictive-texting-takes-toll-childs-brain.html

Harris, S. (2011, July 18). Too much internet use 'can damage teenagers' brains'. *Mail Online*. Retrieved October 23, 2013, from http://www.dailymail.co.uk/sciencetech/article-2015196/Too-internet-use-damage-teenagers-brains.html

Johnson, J. G., Cohen, P., Smailes, E. M., Kasen, S., & Brooke, J. S. (2002). Television viewing and aggressive behavior during adolescence and adulthood. *Science, 295*(5564), 2468-2471.

McVeigh, T. (2001, August 18). Computer games stunt teen brains. *The Guardian* . Retrieved September 10, 2013, from http://www.theguardian.com/world/2001/aug/19/games.schools

Phillips, S. (n.d.). Teens sleeping with cell phones: a clear and present danger. *PBS*. Retrieved September 10, 2013, from http://www.pbs.org/thisemotionallife/blogs/teens-sleeping-cell-phones-clear-and-present-danger

Rowan, C. (2013, January 7). Neuroscience of video game violence - cells that fire together, wire together. *Moving to Learn*. Retrieved September 11, 2013, from http://movingtolearn.ca/2013/neuroscience-of-video-game-violence-cells-that-fire-together-wire-together

Sukel, K. (2012, January 9). Playing video games may make specific changes to the brain. *The Dana Foundation*. Retrieved September 10, 2013, from http://www.dana.org/news/features/detail.aspx?id=34886

Welsh, J. (2011, August 6). Is constant 'Facebooking' bad for teens?. *LiveScience*. Retrieved September 10, 2013, from http://www.livescience.com/15433-facebook-social-media-effects-teens.html

Welsh, J. (2011, November 15). Brains of excessive gamers similar to addicts . *LiveScience*. Retrieved September 10, 2013, from http://www.livescience.com/17033-gamer-brain-reward-system.html

Avery Bedows

[xxiii] Harris, S. (2011, July 18). Too much internet use 'can damage teenagers' brains'. *Mail Online*. Retrieved October 23, 2013, from http://www.dailymail.co.uk/sciencetech/article-2015196/Too-internet-use-damage-teenagers-brains.html

[xxiv] Welsh, J. (2011, August 6). Is constant 'Facebooking' bad for teens?. *LiveScience*. Retrieved September 10, 2013, from http://www.livescience.com/15433-facebook-social-media-effects-teens.html

[xxv] Phillips, S. (n.d.). Teens sleeping with cell phones: a clear and present danger. *PBS*. Retrieved September 10, 2013, from http://www.pbs.org/thisemotionallife/blogs/teens-sleeping-cell-phones-clear-and-present-danger

[xxvi] Grant, C. (2009, August 11). How predictive texting takes its toll on a child's brain. *Mail Online*. Retrieved September 11, 2013, from http://www.dailymail.co.uk/sciencetech/article-1205578/How-predictive-texting-takes-toll-childs-brain.html

[xxvii] Welsh, J. (2011, November 15). Brains of excessive gamers similar to addicts . *LiveScience*. Retrieved September 10, 2013, from http://www.livescience.com/17033-gamer-brain-reward-system.html

[xxviii] McVeigh, T. (2001, August 18). Computer games stunt teen brains. *The Guardian* . Retrieved September 10, 2013, from http://www.theguardian.com/world/2001/aug/19/games.schools

[xxix] Bongiorno, P. (2011, October 6). Your unhappy brain on television. *Psychology Today*. Retrieved October 28, 2013, from http://www.psychologytoday.com/blog/inner-source/201110/your-unhappy-brain-television

[xxx] Ibid.

[xxxi] Cooper, S. (2010, October 10). 6 shocking ways TV rewires your brain. *Cracked*. Retrieved October 28, 2013, from http://www.cracked.com/article_18856_6-shocking-ways-tv-rewires-your-brain.html

[xxxii] Ibid.

[xxxiii] Ibid.

Sleep: An Adolescent's Worst Nightmare

All teenagers know the feeling of sleep deprivation: sluggish thoughts, irritated mood, and, of course, heavy eyelids. Unfortunately, not getting enough sleep has unpleasant consequences. In the late 1990's, a survey of high-school students was conducted that offered a potentially worrying result: it appeared that as students' hours of sleep dropped, so did their grades.

Sleep refreshes us and prepares our bodies and minds for another day of activity. However, there is an alarming trend in our society: it seems that adolescents are, on average, getting far below the amount of sleep they need to perform to their full abilities.

Understanding Sleep

Sleep is universal. There is not a single person on the planet who can survive without it. But, scientifically speaking, sleep is a tricky concept, because a widely accepted explanation of its actual function has been elusive.

To begin we'll talk about sleep physiologically. The human body tells us when it is time to sleep using a structure in our brains, called the pineal gland that produces a chemical compound known as melatonin. This chemical is what makes us feel tired, and induces sleep. **Melatonin** production is caused by a biological timekeeper, known as circadian rhythms, which are rhythms of cellular activity that are primarily controlled by a biological clock found in the anterior hypothalamus. These rhythms have been shown to occur regardless of environmental input (such as the time of day), but are usually aligned.

Circadian rhythms, to some extent, dictate at what time we fall asleep by telling us when we want to sleep. However, humans can engage in a process known as homeostatic control, which among other things, allows us to choose whether or not we sleep, though it does not usually feel pleasant.

It's widely accepted that sleep is good for you. But what *good* means is a highly debated topic. Research makes it seem most likely that during sleep, the cerebral cortex recovers from the day's activities by consolidating neuronal connections that have been developed.

A very recent study of rats has proposed an entirely new theory about the function of sleep: to wash toxins out of the brain. The researchers partaking in this study observed that while the rats were asleep, the volume of their brain cells reduced by 60%, and cerebral fluid washed through to remove a compound called a beta-amyloid protein.[xxiv] Interestingly enough, this protein is found in high amounts in patients who suffer from Alzheimer's Disease. Should further research confirm these findings, a better understanding of the question, "Why do we sleep?" may come to light, while potentially yielding new information about Alzheimer's Disease.

Although the purpose of sleep tends to be a highly debated topic, the stages of sleep are more widely accepted by the scientific community. The stages are as follows:

Stage 1
This stage happens right when you fall asleep, and constitutes the transition between wakefulness and sleep. Usually, it lasts between 1-5 minutes, and comprises roughly 2-5% of sleep per night.

Stage 2
This stage occupies the bulk of sleep each night (usually between 45-65%). Stage 2 sleep acts as a baseline to which your brain will return after entering Stages 3-5.

Stage 3
Stage 3 sleep usually only lasts 10-30 minutes per night. What distinguishes this type of sleep is that during this stage, electrical impulses in the brain slow. During stage 3, the brain only undergoes 1-2 electrical pulse cycles per second. However, interestingly enough, the amplitude of this activity (think, strength of the signal) increases.

Stage 4 and Stage 5
These two very similar stages of sleep constitute what is widely known as REM sleep (the name REM, or Rapid Eye Movement, is given due to the darting back and forth of your eyes during this period of sleep). These phases of sleep, combined, are usually experienced for 20-25% of the night. At the time these phases occur, breathing, heart rate, and brain wave activity become much faster, and it is during this phase of sleep that dreaming occurs. Additionally, the sleeper cannot use their skeletal muscles during this period, meaning they are effectively paralyzed from the neck down.

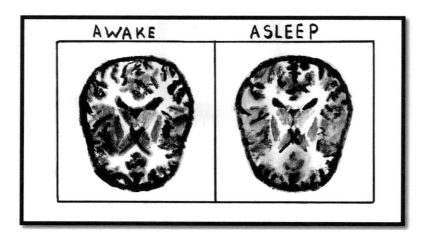

Although how these sleep phases the benefits to our brain take place is still not clear, we can agree that getting enough sleep feels good, and that being tired seems to affect our mood. What, exactly, happens when we don't get sleep?

The Menace of Sleep Deprivation
Trivia question: How much sleep are adolescents supposed to get each night?

A little more than 9 hours.[xxxv]

Another trivia question: How many adolescents get this much sleep per night, on average?

Less than 15%.

Sleep deprivation is *incredibly* common amongst the adolescent population. Less than 15% of teenagers get at least 8.5 hours of sleep on the average night, and more than 25% get fewer than 6.5 hours.[xxxvi] That's a tremendous amount of sleep loss, and when it's compounded across the duration of a school year, an enormous sleep deficit is created.

The question of why adolescents don't get as much sleep as they should seems to have a 3-part answer. The first part of the answer is devastatingly simple: school. Being in high-school brings a tremendous amount of work that leaves little time for sleep. Coupled with after school extracurricular activities, heavy course loads can knock quite a few hours off of the requisite 9.

The second part of the answer to this question may come in the form of something known as a sleep-phase delay. When adolescents hit puberty, their circadian rhythms, which we discussed above, shift slightly. This shift causes teenagers' melatonin production to begin later, meaning the times they naturally want to fall asleep and wake up are pushed later as well. This does not mesh well with school start times at all.

The final piece is slightly more controversial, and could have some serious implications. A study recently done, which is just one in a slew of related studies, has shown that being exposed to electronics with a backlit display for 2 or more hours causes melatonin suppression...by a massive 22%.[xxxvi] Adolescents' exposure to short-wavelength (blue) light at night seems to biologically prevent the brain from producing as much melatonin. As the use of technology for homework and entertainment becomes more and more prevalent, the amount of exposure adolescents have to this blue light before bedtime increases as well. If you have been watching TV for two hours, you will be more awake and consequently want to go to bed later. It is believed that this is a large contributor to why teenagers don't get as much sleep as they should.

Staying up too late can have immediate effects on the function of your brain. Generally, a lack of sleep is associated with greater irritability, decrease in alertness, behavioral problems, and a drop in academic performance. A review[xxxviii] of a series of studies conducted on the subject has yielded seven different types of brain functioning that are impaired by sleep deprivation:

> 1. Difficult situations in which there is an input of unnecessary information that must be sifted through (trying to find the important facts within a dense, shoddily written textbook)
> 2. Strategic planning that involves implementing current/recently occurred events into the thought process
> 3. Creative thinking
> 4. Cost-benefit analysis

5. Interest in rewarding activities
6. Self-assessment
7. Communicating clearly

Although the neurobiological reasons for some of these declines in performance are unknown, there is one section (cost-benefit analysis, which in other words means weighing pros and cons) that scientists have a pretty good idea about. Cost-benefit analysis is a process that takes part in the frontal part of the cerebral cortex, a location in the brain that is involved with concern for risks/negative outcomes in the future. Because sleep deprivation seems to impair the function of this part of the brain, it is possible that sleep helps to facilitate the proper functioning of these structures.

Chronic sleep deprivation has some longterm effects as well. First of all, when teenagers don't get enough sleep for an extended period of time, it seems that their academic performance will suffer. A 1998 study was conducted of high school students, which demonstrated that students who received C's, D's, and F's on their report cards went to bed, on average, 40 minutes later and received 25 minutes less sleep than did the students who received A's and B's.[xxxix]

Another issue with chronic sleep deprivation is the disruption of the circadian rhythm and melatonin production. Longterm suppression of the urge to sleep, and consequently disruption of the circadian rhythm, has been associated with the onset of depression. There is a growing body of evidence that suggests sleep issues predict the onset of this disease. Additionally, when melatonin is suppressed frequently for a long period of time, it has been shown that there is an increase in the chance of suffering from breast cancer. This highlights the need for lesser workloads for teens, so that they no longer have to stay awake for hours in order to finish everything they have to do.

An interesting study was recently conducted that looked at the connection between reward systems in the brain and sleep. This study found that lower amounts/poorer qualities of sleep seem to correlate with the brain's reward system not reacting as strongly to rewarding stimuli.[xl] Although these results do not indicate which factor/observation caused the other, they are interesting because they shed light on possible connections between sleep deprivation and risky behavior. Risky behavior is often thought to come from getting a lower sense of satisfaction from activities that ordinarily would be pleasing. Teenagers are notorious for being risk-takers, and this evidence could lead way for a new understanding of why.

In Conclusion
Sleep seems to play a tremendous role in affecting how we function on a day-to-day basis, and can even have implications for down the road. For adolescents in particular, getting enough sleep seems to be extremely important to performing well academically, and helps avoid depression, a disease that affects many teenagers.

While the jury is still out on *how* exactly sleep causes its positive benefits, what is for sure is that it helps maintain our brains' abilities to function. One such function that sleep has an effect on is memory. In the next chapter, we'll take a look at the fascinating topic of human recall.

Avery Bedows

Terms:

Melatonin -
Melatonin is a chemical produced in the front of the hypothalamus whose job is to prepare the body for sleep. It creates the sensation of "tiredness," and allows the body to transition from wakefulness to sleep.

Circadian Rhythms -
Circadian rhythms are a cycle of biological activity that occur in a specific, repetitive pattern over a long duration of time. The rhythms influence the times at which melatonin production increases.

Homeostatic Control -
Homeostatic control, when used in this context, describes humans' ability to control their homeostasis (state of internal equilibrium) to a certain degree. Being able to stay awake despite being tired is an example of the control humans can consciously exercise over their involuntary homeostatic functions.

References
Breecher, M. M. (n.d.). The biology of dreaming. *Columbia University in the City of New York*. Retrieved October 21, 2013, from http://www.columbia.edu/cu/21stC/issue-3.4/breecher.html

Carpenter, S. (2000). Sleep deprivation may be undermining teen health. *American Psychological Association*. Retrieved September 15, 2013, from http://www.apa.org/monitor/oct01/sleepteen.aspx

Dawson, Peg. (2005). Sleep and adolescents. *National Association of School Psychologists*. Retrieved September 15, 2013 from http://www.nasponline.org/resources/principals/sleep%20disorders%20web.pdf

Different stages of sleep. (n.d.). *ThinkQuest*. Retrieved October 21, 2013, from http://library.thinkquest.org/C005545/english/sleep/stage.htm

Harrison, Y., & Horne, J. A. (2000). The impact of sleep deprivation on decision making: a review. *Journal of Experimental Psychology: Applied, 6*(3), 236-249. Retrieved September 15, 2013, from http://dx.doi.org/10.1037//1076-898X.6.3.236

Holm, S. M., Forbes, E. E., Ryan, N. D., Phillips, M. L., Tarr, J. A., & Dahl, R. E. (2009). Reward-related brain function and sleep in pre/early pubertal and mid/late pubertal adolescents. *Journal of Adolescent Health, 45*(4), 326-334.

Kim, M. (2013, October 19). Brains flush toxic waste in sleep, including Alzheimer's-linked protein, study of mice finds. *Washington Post*. Retrieved October 23, 2013, from http://www.washingtonpost.com/national/health-science/brains-flush-toxic-waste-in-sleep-including-alzheimers-linked-protein-study-of-mice-finds/2013/10/19/9af49e40-377a-11e3-8a0e-4e2cf80831fc_story.html

Mullaney, R. (n.d.). Light from self-luminous tablet computers can affect evening melatonin, delaying sleep. *Rensselaer Polytechnic Institute*. Retrieved October 21, 2013, from http://news.rpi.edu/luwakkey/3074?destination=node/145

Nauert, R. (n.d.). Sleep restores brain energy. *Psych Central*. Retrieved October 21, 2013, from http://psychcentral.com/news/2010/06/30/sleep-restores-brain-energy/15240.html

Vitaterna, M. H., Takahashi, J. S., & Turek, F. W. (n.d.). Overview of circadian rhythms. *NIAAA Publications*. Retrieved October 21, 2013, from http://pubs.niaaa.nih.gov/publications/arh25-2/85-93.htm

Avery Bedows

[xxxiv] Kim, M. (2013, October 19). Brains flush toxic waste in sleep, including Alzheimer's-linked protein, study of mice finds. *Washington Post*. Retrieved October 23, 2013, from http://www.washingtonpost.com/national/health-science/brains-flush-toxic-waste-in-sleep-including-alzheimers-linked-protein-study-of-mice-finds/2013/10/19/9af49e40-377a-11e3-8a0e-4e2cf80831fc_story.html
[xxxv] Dawson, Peg. (2005). Sleep and adolescents. National Association of School Psychologists. Retrieved September 15, 2013 from http://www.nasponline.org/resources/principals/sleep%20disorders%20web.pdf
[xxxvi] Ibid.
[xxxvii] Mullaney, R. (n.d.). Light from self-luminous tablet computers can affect evening melatonin, delaying sleep. *Rensselaer Polytechnic Institute*. Retrieved October 21, 2013, from http://news.rpi.edu/luwakkey/3074?destination=node/145
[xxxviii] Harrison, Y., & Horne, J. A. (2000). The impact of sleep deprivation on decision making: a review. *Journal of Experimental Psychology: Applied*, *6*(3), 236-249. Retrieved September 15, 2013, from http://dx.doi.org/10.1037//1076-898X.6.3.236
[xxxix] Carpenter, S. (2000). Sleep deprivation may be undermining teen health. *American Psychological Association*. Retrieved September 15, 2013, from http://www.apa.org/monitor/oct01/sleepteen.aspx
[xl] Holm, S. M., Forbes, E. E., Ryan, N. D., Phillips, M. L., Tarr, J. A., & Dahl, R. E. (2009). Reward-related brain function and sleep in pre/early pubertal and mid/late pubertal adolescents. *Journal of Adolescent Health*, *45*(4), 326-334.

Memory: Easier to Improve than You Think

Memory is one of, if not the most essential cognitive function. All types of learning are a form of memory, so without it, you would never move beyond the mentality of a baby. For adolescents, who are constantly learning, through school, social interactions, and day-to-day life, memory is exceptionally important.

Here are some common preconceptions you might have about how memory functions:

1. Human memory operates like a memory card, with all of the information stored statically in one place.

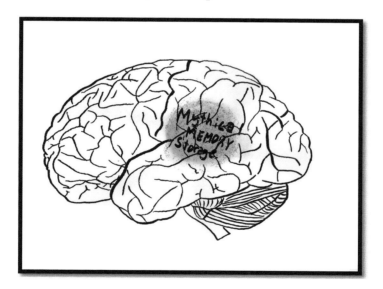

2. When you remember something, you are remembering one, individual thing.
3. Memory involves brain growth.

These happen to all be very, very incorrect. Let's fix those mis(pre)conceptions.

The first one: memories are actually distributed and stored in different regions. There is no central "memory card" that keeps all of the information.

The second one: memories are split up into tiny pieces and stored in many different locations in the brain. This means that when you try and recall something, it's not a single memory being pulled from one location. Rather, it's segments of a memory being pieced together!

The third one: when you remember something, it doesn't mean that more cells magically pop up in your brain. In actuality, more synapses are being created.

Adolescents and Memory

The difference in brain anatomy between adolescents and adults causes adolescents to learn differently. The first difference that impacts memory is in the volume of white matter. White matter, which consists of neurons that have been myelinated (see previous chapters for a definition of myelination), is responsible for the transfer of information between cortical lobes, as well as to and from the spinal cord.

Myelinated neurons increase the transmission speed of nerve impulses, which carry information. Adolescent brains are not fully myelinated, causing information to travel less quickly through the neural network. This decreases the efficiency of memory/learning, making it more difficult for adolescents to learn and understand quickly,

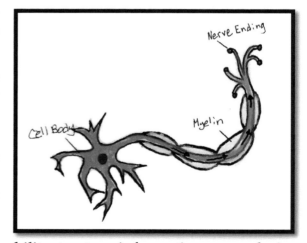

and can even impair their ability to store information properly in their brains. Think about trying to teach a toddler that the Earth is round. They spend ages asking questions, and even then, often forget what you've said. Their brains, with significantly less white matter, take a long time to process and store information.

Short Term Memory

Short term memory, unsurprisingly, constitutes memories that are held only for a short period of time. Sensory memory is the briefest form of memory. It occurs when a copy of sensory information is stored in the brain, and this lasts from anywhere between 0.5 - 2 seconds.[xi] Short term memory, however, refers to the brain's ability to retain around 7 pieces of information for a period of between 15-30 seconds.ˣˡⁱⁱ You can extend this time by repeating this information over and over to yourself. This is called "rote rehearsal."

Short term memory is important because it allows you to remember and process information simultaneously. For example, when you're reading this sentence, you're both remembering the beginning (which you have already processed), and processing the end.

The prefrontal cortex (PFC) is considered important for holding on to short term memories. It does not, however, "store" the information per-se. Scientists have determined the connection between the prefrontal cortex and short term memory by observing that people with damaged prefrontal cortices tend to have impaired short term memory. this indicates there is a connection between the PFC and short term memory, but does not tell us that the memories are *stored* in the PFC.

Additionally, the prefrontal cortex is where short term memory and long term memory come together and interact. The prefrontal cortex extracts long term memories and combine them with short term memories – an element that is one of the fundamental basics of thinking. This process is called "retrieval."

Long Term Memory
Long term memory, as its name suggests, constitutes the memories that can be held on to for longer than about 30 seconds. The process of retaining these memories utilizes synaptic growth – the changing of neuronal connectivity. Long term memory is thought to have two systems: **declarative** and **nondeclarative**. These systems, which will be explained below, are thought to be competitive; this means that if one system is not adequate, the other is ready to take over.

<u>Declarative Memory</u>: Declarative memory is long term memory for facts and events, and these memories are consciously accessible (you know you have them, or can *declare* you have them). For example, reciting the U.S. Constitution's Preamble pulls from a declarative memory; so does remembering your mother's birthday. This form of memory heavily relies on the hippocampus – not for storage, but as a direction-giver. Information from sensory areas of

the brain converges at or around the hippocampus, and there it gets strung into one large, encompassing memory.

These memories are stored, neurally, as synaptic changes. It is thought that the hippocampus helps replicate the pattern of neuron activation from the original experience, which is what solidifies the memories. Old memories are simply the product of multiple reiterations of the same neuron firing pattern. This points to declarative memories (otherwise known as explicit memories) being stored in the systems used originally to process them. Visible memories would be stored in the occipital lobes; sounds in the temporal lobes; etc.

Nondeclarative Memory: Nondeclarative memory is a long term memory that falls under one of three categories: procedural, priming, or conditioning. Procedural memory is when you learn a process to the point where it becomes habitual. It is thought that the striatum, the brain's reward system, helps to orchestrate this type of memory by reinforcing processes through positive reinforcement. Priming is when you improve at something after repeating it multiple times, and is thought to occur because of a change in how stimulus is dealt with cognitively – the more you do it, the better your brain gets at doing it. This process most likely involves the neocortex (the outer layer of the cortex). Finally, there are two types of conditioning: emotional response and muscle memory. Nondeclarative emotional responses are memories that respond emotionally to certain stimuli. If a situation originally evoked a certain emotion, then after repetition, the same type of situation will create a very similar emotional response. For instance, if you had a particularly bad experience at your middle school (as many people seem to), going back there later on in life might upset you. The amygdala is believed to be heavily involved in this process. Muscle memory, which constitutes things such as knowing how to properly throw a baseball or perform scales on a piano, depends on the cerebellum to function properly.

How to Optimize Your Memory

You can use certain techniques to make remembering information significantly less difficult. These can be implemented when you're studying, or simply whenever you want to remember things more easily.

The first method is something called "chunking." Because your short-term memory can store only 7 items, you can chunk groups of numbers, letters, etc. together into one big group that you already recognize so that you can hold onto many more individual items. For instance, let's take a series of 9 letters: C N B A T B C C B. Storing these individually is very difficult. But, if you rearrange the letters, they can also be stored as NBC CAT BBC. All three of these grouped letters form recognizable words/acronyms. The 9 items have been transformed into 3 easily rememberable items! These three items are also much easier to bring into long term memory than the 9. The ideal chunk size is roughly 3 pieces of information, and because each chunk only counts as one item, you can store just as many chunks as you would normal items as a short term memory (i.e., 7 on average). That means you can store around 20 pieces of information in your short term memory using chunking!

You can also use metaphors and analogies to help remember. Connecting information you are trying to learn with information you have already stored in your brain creates associations that assist you with memory. This would be a great way to remember how to do something in school. If you know another process that's similar, make the connections between the new one and the old one. For example, when learning division, it would help to connect it to your current knowledge of multiplication as they are similar processes. Chances are, you'll have a much easier time recalling it.

The presence of humor also helps memory. Humor is thought to be highly involved with the amygdala, which is one of the structures that is relied upon for memory, especially for adolescents. Using humor is an ideal way to connect material with emotion, assisting the memory process. If you can somehow make American History, or whatever topic you are studying, funny, then you will be more emotionally engaged and your amygdala will be in use, helping with your recall later.

Learning/memory formation is severely hindered by fear. In a fearful environment, nearly no learning occurs. It is important to be in a comfortable situation when learning, otherwise the brain is more focused on survival than holding on to the current stimuli. It's no wonder most adolescents do much better in classes with friendly teachers.

Similarly, whether or not something is pleasant impacts how well you will remember it. Vivid, colorful, sensory-input-filled images can help you remember something. However, it's best to make these images pleasant ones, because sometimes the brain blocks out unpleasant things, which is a good survival strategy on your brain's part, but not good for your memory.

In Conclusion

Memory is an essential feature that lets us function in our surroundings. It allows us to think, to consciously draw on past experiences and use them to formulate new ideas and to problem solve. Memory also helps us to establish our place in society by being able to keep tracks of things past and things present so that we have a reference frame.

Memories are powerful, and can remain consciously accessible for a lifetime. In the next chapter, we will discuss certain mental diseases which may in part be caused by experiences that cause traumatic memories.

Avery Bedows

<u>Terms</u>:

Short Term Memory -
Short term memories are memories that are stored by the brain for anywhere between 15-30 seconds.

Long Term Memory -
Long term memories are memories that have been stored in the parts of the brain that originally experienced them, and can be accessed long after they were initially introduced to the brain.

Declarative -
Declarative memories are facts or events that you can consciously remember.

Nondeclarative -
Nondeclarative memories are processes, skills, habits, or types of conditioning that you cannot consciously access.

References

Byrne, J. H. (n.d.). Learning and memory. *Neuroscience Online.* Retrieved September 14, 2013, from http://neuroscience.uth.tmc.edu/s4/chapter07.html

Introduction to memory techniques. (n.d.). *Mind Tools.* Retrieved January 5, 2014, from http://www.mindtools.com/memory.html#sthash.Ze14RHex.dpuf

Knowlton, B. J., & Foerde, K. (2008). Neural representations of nondeclarative memories. *Current Directions in Psychological Science, 17*(2), 107-111. Retrieved December 3, 2013, from http://dx.doi.org/10.1111/j.1467-8721.2008.00558.x

Krupa, A. K. (2009). The competitive nature of declarative and nondeclarative memory systems: converging evidence from animal and human brain studies. *UCLA Undergraduate Science Journal, 22.* Retrieved December 3, 2013, from http://www.studentgroups.ucla.edu/USJ/The_Competitive_Nature_of_Declarative_and_Nondeclarative_Memory_Systems_AKKrupa.pdf [PDF Form]

Lorain, P. (n.d.). Brain development in young adolescents . *National Educational Association.* Retrieved January 5, 2014, from http://www.nea.org/tools/16653.htm

Mastin, L. (2010). Short-term memory and working memory. *The Human Memory.* Retrieved January 5, 2014, from http://www.human-memory.net/types_short.html

Memory and Cognition . (n.d.). *Neuroscience Resource Page.* Retrieved December 3, 2013, from http://www.neuroanatomy.wisc.edu/coursebook/neuro6(2).pdf [PDF Form]

Mohs, R. C. (2007, May 8). How human memory works. *HowStuffWorks.* Retrieved November 5, 2013, from http://science.howstuffworks.com/life/inside-the-mind/human-brain/human-memory2.htm

---

[xli] Memory and Cognition . (n.d.). *Neuroscience Resource Page.* Retrieved December 3, 2013, from http://www.neuroanatomy.wisc.edu/coursebook/neuro6(2).pdf [PDF Form]
[xlii] Ibid.

Troubleshooting Your Brain

Many people view mental disorders as something that only "crazy" people have. The tragic truth is that they're surprisingly common. You aren't abnormal at all if you have a disorder. Here are some of the sobering statistics about the prevalence of certain mental disorders in children and adolescents:

Depression
Annually, 2% of young children, 4% of young adolescents, and at least 16% of older adolescents are depressed.[xliii]

Anxiety
Between 15-20% of children have some sort of anxiety disorder.[xliv]

Eating Disorders
14-22% of children or adolescents have an eating disorder.[xlv]

Attention Deficit Hyperactivity Disorder (ADHD)
In the United States, 5-8% of children have been diagnosed with ADHD.[xlvi]

Oftentimes, mental disorders are viewed as a 'fake' illness, or a weakness of personality. This is incorrect: mental disorders are biologically explicable, and their effects are just as tangible and accepted within the medical community as any other physical disorder.

Depression in Adolescents
Depression is extremely prevalent in adolescents: at least 3 out of every 20 teenagers will experience it each year. This mental disorder has a variety of negative symptoms, making it hard to perform regular activities.

Depression is a disorder in which the patient experiences intense sadness or numbness; some describe it as feeling like a grey shroud over all of their senses. This can manifest in any of a number of ways. First of all, depression can cause dysregulated sleep. Someone who has this symptom may experience insomnia (inability to sleep), hypersomnia (sleeping too much), or fatigue (exhaustion). Additionally, depressed patients may notice irregular eating patterns, such as appetite changes or weight gains/losses. Furthermore, depression affects motor behavior. A slowing-down of thoughts and physical activity, called psychomotor retardation, can occur. Additionally, psychomotor agitation and restlessness, which are involuntary, repetitive actions, can become present as well. These may include biting nails or chewing the inside of your cheek, and as a result can be harmful.

Depression also takes a toll on cognition. Those who are depressed may notice an increase in distractibility, a decrease in attention span, a decrease in their ability to make decisions, and a rising sense of hopelessness. Also, self-esteem can be shattered. Many people experience feelings of guilt and worthlessness, and begin to contemplate suicide along with other forms of self-harm.

Depression is a diagnosable medical condition. Generally, depressive symptoms will be diagnosed in one of two ways. Less severe, longer-lasting symptoms usually indicate **dysthymic disorder**, which is essentially a more mild form of depression. Nearly 80% of children with dysthymic disorder will develop full-on depression later on,[xlvii] which is perhaps an indication of a lack of effective treatments. More severe symptoms will be diagnosed as depression, ranging from mild depression to **Major Depressive Disorder**, which is classified by chronic and severe depressive symptoms. There are a variety of treatments used to address these disorders, including antidepressant medications, psychotherapy, and a variety of other treatments whose efficacies are still being evaluated.

Research is still being conducted about *why* exactly depression occurs. Certain risk factors have been determined to influence the chances someone has of becoming depressed. Environmental factors such as abuse (physical, mental, or sexual), traumatic social situations, or deaths in the family have been shown to increase the probability of becoming depressed. Genetic factors are being explored as well; specifically, the suppression of the BDNF gene, which controls the presence of BDNF in the brain, may lead to depression. It is possible that those who have genetic mutations to this gene might be more likely to suffer from depression than those who do not.

Depression is visible in the brain. Using MRI scans, the brains of depressed patients can be evaluated by comparing them to the brains of patients who do not suffer from this disorder. It appears that depression is correlated with decreased electrical activity in the brain. It has been demonstrated that there is less activity in the parietal lobes of people suffering from depression, which may have to do with emotional expression and perception.[xlviii] Additionally, brain scans have shown reduced volume in the frontal lobes, frontal white matter, and the left orbitofrontal cortex in depressed patients.[xlix]

Anxiety and Adolescents

Anxiety is the brain's response to events and stimuli that it considers dangerous, and can involve fear, nervousness, and among other things, a queasy stomach. Certain people have disorders in which this response is hyperactive, making them feel excessively anxious with little aggravation. There are several different types of anxiety: separation anxiety, social anxiety, Post Traumatic Stress Disorder (PTSD), general phobias, General Anxiety Disorder (GAD), and others.

Although these different types of anxiety involve different triggers, their physiological and emotional consequences are more-or-less the same. The physical symptoms of anxiety are sweating, increased heart rate, and gastrointestinal agitation. Emotionally, sufferers of these disorders experience fear, unexplainable terror, and nervousness.

Roughly 15-20% of people experience some sort of anxiety disorder throughout their lifetime.[l] Females typically experience higher levels of stress than males do, and consequently have higher rates of anxiety. Genetics influence anxiety as well, as its genetic heritability (what fraction of the population's anxiety is transmitted form a parent) is, according to an average of various studies, 77%.[ii]

Similar to depression, the presence of anxiety is visible in the brain. Brain scans have shown that patients who suffer from an anxiety disorder demonstrate an increased amygdala response when shown fearful faces.[iii] This observation seems to occur in general: the amygdala of anxious patients are hypersensitive to stimuli, meaning people suffering form anxiety can become more stressed out with less cause than the average person. Interestingly, the amygdala and other parts of the emotional brain (limbic system) have been shown to activate when females with anxiety are evaluating other females.[iii]

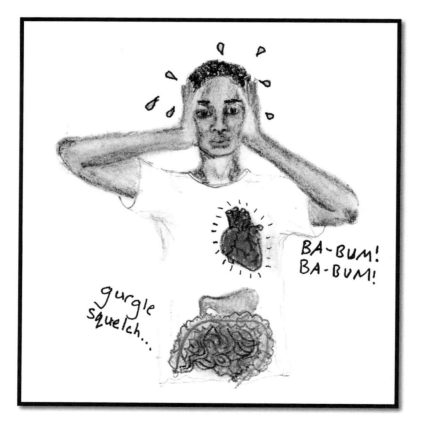

Anxiety seems to lead to other mental disorders. Specifically, adolescent anxiety has been associated with the later development of Major Depressive Disorder. The greater quantity/more severe anxiety disorders you have, the more likely you are to experience depression at some point in your life.

Adolescents and Eating Disorders
There are three main types of eating disorders: **Anorexia Nervosa,
Bulimia Nervosa,** and **Binge Eating Disorder.** Patients who suffer
from Anorexia often have an underlying emotional concern with
body image, and as a result may starve themselves or exercise
excessively. Those who suffer from Bulimia have similar
underlying concerns, but will oftentimes eat too much and then
force themselves to regurgitate their meals afterwards, or will
consume laxatives so that the food they just ate cannot be absorbed
by their bodies. People with Binge Eating Disorder feel
uncontrollable impulses to eat excessively.

Eating disorders are often said to be a "women's issue". However,
this is not 100% true. A recent study on a university campus found
that positive screens for eating disorders had a female-to-male ratio
of 3-to-1.[iv] Both females and males experience societal pressure to
look a certain way from a young age. For young girls, Barbie,
impossibly skinny with long legs, is the archetype of a woman and
for young boys, an action figure, massively muscular while
maintaining a tiny waist, is the archetype of a man. From what is
presented to children as perfection, it should not be surprising that
81% of ten-year old girls are afraid of being fat.[iv]

Anorexia Nervosa is a disease in which people starve themselves,
exercise too much, or abuse laxatives. Oftentimes, those with
Anorexia may appear frail, be constantly fatigued but suffer from
insomnia, and have thinning hair and dry skin. Additionally, they
may have irregularities in their heart rhythm, low blood pressure,
and may be prone to dehydration. People with anorexia often
demonstrate lack of emotion, social seclusion, increased irritability,
and in many instances, depression. Girls with Anorexia may not
menstruate.

A frequent, and incorrect, preconception about Anorexia is that it is
always about looking thin. Sometimes, when someone, especially
someone young, feels that they do not have adequate control of
their life, they resort to controlling what goes in and out of their
body. Anorexia can be very dangerous when it manifests in this
form because it can often seem like a slightly abnormal obsession
with health, instead of a disorder.

The causes of Anorexia Nervosa are uncertain. Two possibilities are
genetic factors and issues with serotonin, similar to depression.

Additionally, people with obsessive-compulsive traits are more likely to acquire this disease. Additionally, many believe that the emphasis placed by western culture, particularly in the media, on being thin increases the occurrence of Anorexia in our society.

There are a few important risk factors for Anorexia. First of all, it is a disease that tends to be more common in teenagers than in adults. Additionally, it seems to be impacted by family history: if you have someone in your immediate family who has or has had Anorexia, you have a much higher chance of developing it, both from learning behaviors from that family member, and the shared genes. Furthermore, undergoing periods of difficult transition can heighten the risk of anorexia.

Anorexia, in addition to causing physiological problems such as anemia, bone loss, gastrointestinal issues, and kidney malfunction, can also cause other mental disorders. Anorexia has been associated with depression, anxiety, personality disorders, obsessive-compulsive disorder, and drug abuse.

Anorexia can be fatal. Because of the various serious health effects that it has, treatment for this disorder might include the monitoring of vital signs, or even hospitalization. Additionally, the use of psychotherapy and antidepressant medication is put in practice to address underlying/other simultaneously existing conditions which may be influencing the Anorexia.

Unfortunately, what many people do not realize how deadly this eating disorder is. In fact, Anorexia Nervosa has the highest mortality rate of any mental illness: 5.9%.[lvi]

Bulimia Nervosa is also dangerous. There are two different types of Bulimia: purging Bulimia, and nonpurging Bulimia. Someone who has purging Bulimia may eat excessive amounts of food and then make themselves vomit, or use laxatives or diuretics to keep off weight. Nonpurging Bulimia involves the use of fasting, dieting, or an excess of exercise to control weight. Like Anorexia, Bulimia is a condition whose underlying causes are closely related to self-image.

Although there is no consensus from a biological standpoint about what causes Bulimia, certain risk factors have been identified. The first one is being female. Females seem significantly more likely to be bulimic than males do. Additionally, adolescents are more likely to acquire this disorder than adults are. Similar to Anorexia, having a first-degree relative (parent or sibling) who suffers from Bulimia

makes you significantly more likely to have it at some point. Also in a similar manner to Anorexia, pressure from western culture and the media about body image is believed to be, at least in-part, culpable for the presence of Bulimia in our society.

Bulimia can cause severe dehydration, digestion issues, anxiety and depression. Treatments for this eating disorder include various types of psychotherapy, or a combination of psychotherapy with antidepressant medications.

Finally, Binge Eating Disorder (BED) is when one eats too much food too often because of compulsions that are very difficult – nearing on impossible – to control. People who suffer from BED go on uncontrollable eating binges to the point of being painfully full, and often have feelings of disgust or depression towards this bingeing afterwards.

The causes of BED are unknown, but family history, underlying psychological issues, and age (BED mostly occurs in people in their late teens through their early 20s) all are important risk factors. Additionally, a history of frequent dieting may increase the chances of getting this disorder.

BED is extremely dangerous. Not only does it lead to insomnia and depression, but people with BED often become highly obese, have high blood pressure, type 2 diabetes, high cholesterol, heart disease, joint and muscle pain, and headaches. Treatments for this disorder include psychotherapy, antidepressants, and anticonvulsant medications.

ADHD in Adolescents
Attention Deficit Hyperactivity Disorder is a behavioral disorder that impairs people's ability to focus. This disorder usually begins affecting children at an early age, is quite prevalent, and is the most frequently treated disorder in children.[lvii]

Attention Deficit Disorder (ADD) is very similar to ADHD. The only difference is that patients with this disorder do not experience the symptoms of hyperactivity. Because the other symptoms, causations, and risk factors remain the same for both of the disorders, we will focus only on ADHD.

There are numerous symptoms of ADHD that can seriously detract from an adolescent's ability to learn. The symptoms fall under three main categories: inattention, hyperactivity, and impulsivity. The specific symptoms of inattention are the ability to be distracted easily, having difficulty focusing, having difficulty following directions, and having difficulty processing information quickly. Hyperactivity includes fidgeting and squirming, as well as excessive talking. Impulsivity includes impatience, and engaging in actions that show a disregard for consequences.

The tricky thing about ADHD is that there is no cure. Although there are certain medications whose controversial use and effects seem to increase the ability of children and adolescents with ADHD to focus, the effects of these medications are temporary. One other option is behavioral therapy, which works to change the way in which a patient behaves, hopefully reducing the symptoms of ADHD over time.

Not much is known about the causes of ADHD. However, some studies have indicated that ADHD is influenced heavily by family incidence (genetics), and that when mothers smoke cigarettes and consume alcohol during pregnancy, their children have a higher risk of having ADHD.[lviii]

There are a few things that go on in the brain which are thought to influence ADHD. Studies have shown that the dorsolateral prefrontal cortex, which is associated with organization, planning, and working memory, along with attention, does not function properly in patients with ADHD.[ix] Additionally, lesions in the orbitofrontal cortex have been associated with social inhibitions and behavioral disorders, which indicates that this section of the brain may be involved with this disorder. Furthermore, it has been demonstrated in numerous studies that patients diagnosed with ADHD, when performing tasks that involve controlling their inhibitions, have less activity in their dorsal anterior cingulate cortex than patients who do not.[ix] A series of imaging studies showed that children/adolescents with ADHD showed cortical maturation delayed by 3 years, and that the corpus callosum, which connects the two hemispheres of the brain, has an abnormal pattern of growth.[lxi]

In Conclusion

Mental disorders are conditions that many people have to deal with. They should not be dismissed, as they all-to-often are. If you or anyone you know is experiencing symptoms that sound like any of the disorders talked about in this chapter, it is extremely important that you seek professional assistance. These disorders are manageable, but it is imperative they be addressed properly.

When mental disorders such as depression are not taken care of, people tend to turn to unsafe means of dealing with the trauma. One such unsafe method, which is talked about in the next chapter, is the abuse of alcohol and other drugs.

Avery Bedows

<u>Terms:</u>

Depression -
This is a mental disorder characterized by extreme sadness, disinterest in daily activities, and in severe cases, suicidal ideation.

Dysthymic Disorder -
This is a disorder whose milder, longer-lasting symptoms mimic those of depression.

Major Depressive Disorder -
This is a classification of depression with symptoms of a certain degree of severity and chronicity.

Anxiety -
This is the name of a broad series of disorders that result in paralyzing fear that can have physiological effects.

Anorexia Nervosa -
This is a disorder in which concern for bodyweight causes someone to stop eating/eat far too little to be properly nourished.

Bulimia Nervosa -
This mental disorder is when people force themselves to regurgitate what they eat so they will not put on weight.

Binge Eating Disorder -
Binge Eating Disorder is a disorder in which people eat excessively.

Attention Deficit Hyperactivity Disorder -
This is a mental disorder that affects your ability to focus properly.

References

Anorexia nervosa. (2012, January 5). *Mayo Clinic*. Retrieved November 11, 2013, from http://www.mayoclinic.com/health/anorexia/DS00606

Anxiety disorders in children and adolescents (fact sheet). (n.d.). *National Institute of Mental Health*. Retrieved September 15, 2013, from http://www.nimh.nih.gov/health/publications/anxiety-disorders-in-children-and-adolescents/index.shtml

Beesdo, K., Knappe, S., & Pine, D. S. (2009). Anxiety and anxiety disorders in children and adolescents: developmental issues and implications for DSM-V. *Psychiatric Clinics of North America*, 32(3), 483-524. Retrieved September 15, 2013, from http://dx.doi.org/10.1016/j.psc.2009.06.002

Binge-eating disorder. (2012, April 3). *Mayo Clinic*. Retrieved November 11, 2013, from http://www.mayoclinic.com/health/binge-eating-disorder/DS00608

Brain imaging studies reveal neurobiology of eating disorders. (2013, April 10). *ScienceDaily*. Retrieved September 15, 2013, from http://www.sciencedaily.com/releases/2013/04/130410191559.htm

Bulimia nervosa. (2012, April 3). *Mayo Clinic*. Retrieved November 11, 2013, from http://www.mayoclinic.com/health/bulimia/DS00607

Mazzone, L., Ducci, F., Scoto, M. C., Passaniti, E., D'Arrigo, V., & Vitiello, B. (2007). The role of anxiety symptoms in school performance in a community sample of children and adolescents. *BMC Public Health*, 7(1), 347. Retrieved September 15, 2013, from http://dx.doi.org/10.1186/1471-2458-7-347

Miller, A. (2007). Social neuroscience of child and adolescent depression. *Brain Cognition*, 65(1), 47-58. Retrieved September 15, 2013, from http://dx.doi.org/10.1016/j.bandc.2006.02.008

(1995). Morbidity in anorexia nervosa. *American Journal of Psychiatry*, 152(7), 1073-1074. Retrieved November 8, 2013, from http://ajp.psychiatryonline.org/article.aspx?articleID=171119

Morgan, J. K., Shaw, D. S., & Forbes, E. E. (2013). Psychological and behavioral engagement in social contexts as predictors of adolescent depressive symptoms. *Journal of Youth and Adolescence*, 42(8), 1117-1127. Retrieved September 15, 2013, from http://dx.doi.org/10.1007/s10964-012-9815-2

Spencer, T. J., Biederman, J., & Mick, E. (2007). Attention-deficit/hyperactivity disorder: diagnosis, lifespan, comorbidities, and neurobiology. *Journal of Pediatric Psychology*, 32(6), 631-642. Retrieved September 15, 2013, from http://dx.doi.org/10.1093/jpepsy/jsm005

Statistics on males and eating disorders. (n.d.). *National Eating Disorders Association*. Retrieved August 18, 2014, from http://www.nationaleatingdisorders.org/statistics-males-and-eating-disorders

Swanson, S. A., Crow, S. J., Grange, D. L., Swendsen, J., & Merikangas, K. R. (2011). Prevalence and correlates of eating disorders in adolescents: results from the national comorbidity survey replication adolescent supplement. *JAMA Psychiatry* , 68(7), 714-723. Retrieved September 15, 2013, from http://dx.doi.org/10.1001/archgenpsychiatry.2011.22

The war on women's bodies. (n.d.). *National Eating Disorders Association*. Retrieved August 18, 2014, from http://www.nationaleatingdisorders.org/war-womens-bodies

[xliii] Miller, A. (2007). Social neuroscience of child and adolescent depression. *Brain Cognition*, *65*(1), 47-58. Retrieved September 15, 2013, from http://dx.doi.org/10.1016/j.bandc.2006.02.008

[xliv] Beesdo, K., Knappe, S., & Pine, D. S. (2009). Anxiety and anxiety disorders in children and adolescents: developmental issues and implications for DSM-V. *Psychiatric Clinics of North America*, *32*(3), 483-524. Retrieved September 15, 2013, from http://dx.doi.org/10.1016/j.psc.2009.06.002

[xlv] Swanson, S. A., Crow, S. J., Grange, D. L., Swendsen, J., & Merikangas, K. R. (2011). Prevalence and correlates of eating disorders in adolescents: results from the national comorbidity survey replication adolescent supplement. *JAMA Psychiatry* , *68*(7), 714-723. Retrieved September 15, 2013, from 10.1001/archgenpsychiatry.2011.22

[xlvi] Spencer, T. J., Biederman, J., & Mick, E. (2007). Attention-deficit/hyperactivity disorder: diagnosis, lifespan, comorbidities, and neurobiology. *Journal of Pediatric Psychology*, *32*(6), 631-642. Retrieved September 15, 2013, from http://dx.doi.org/10.1093/jpepsy/jsm005

[xlvii] Miller, A. (2007). Social neuroscience of child and adolescent depression. *Brain Cognition, 65*(1), 47-58. Retrieved September 15, 2013, from http://dx.doi.org/10.1016/j.bandc.2006.02.008

[xlviii] Ibid.

[xlix] Ibid.

[l] Spencer, T. J., Biederman, J., & Mick, E. (2007). Attention-deficit/hyperactivity disorder: diagnosis, lifespan, comorbidities, and neurobiology. *Journal of Pediatric Psychology, 32*(6), 631-642. Retrieved September 15, 2013, from http://dx.doi.org/10.1093/jpepsy/jsm005

[li] Ibid.

[lii] Ibid.

[liii] Anxiety disorders in children and adolescents (fact sheet). (n.d.). *National Institute of Mental Health*. Retrieved September 15, 2013, from http://www.nimh.nih.gov/health/publications/anxiety-disorders-in-children-and-adolescents/index.shtml

[liv] Statistics on males and eating disorders. (n.d.). *National Eating Disorders Association*. Retrieved August 18, 2014, from http://www.nationaleatingdisorders.org/statistics-males-and-eating-disorders

[lv] The war on women's bodies. (n.d.). *National Eating Disorders Association*. Retrieved August 18, 2014, from http://www.nationaleatingdisorders.org/war-womens-bodies

[lvi] (1995). Morbidity in anorexia nervosa. *American Journal of Psychiatry, 152*(7), 1073-1074. Retrieved November 8, 2013, from http://ajp.psychiatryonline.org/article.aspx?articleID=171119

[lvii] Spencer, T. J., Biederman, J., & Mick, E. (2007). Attention-deficit/hyperactivity disorder: diagnosis, lifespan, comorbidities, and neurobiology. *Journal of Pediatric Psychology, 32*(6), 631-642. Retrieved September 15, 2013, from http://dx.doi.org/10.1093/jpepsy/jsm005

[lviii] Anxiety disorders in children and adolescents (fact sheet). (n.d.). *National Institute of Mental Health*. Retrieved September 15, 2013, from http://www.nimh.nih.gov/health/publications/anxiety-disorders-in-children-and-adolescents/index.shtml

[lix] Spencer, T. J., Biederman, J., & Mick, E. (2007). Attention-deficit/hyperactivity disorder: diagnosis, lifespan, comorbidities, and neurobiology. *Journal of Pediatric Psychology, 32*(6), 631-642. Retrieved September 15, 2013, from http://dx.doi.org/10.1093/jpepsy/jsm005

[lx] Ibid.

[lxi] Anxiety disorders in children and adolescents (fact sheet). (n.d.). *National Institute of Mental Health*. Retrieved September 15, 2013, from http://www.nimh.nih.gov/health/publications/anxiety-disorders-in-children-and-adolescents/index.shtml

The Effects of Alcohol, Marijuana, and Prescription Drug Abuse

Drugs and alcohol are undoubtedly bad for your health because they are essentially poisons. But, adults and educators aren't always clear on what exactly makes them bad. The usual method of persuading adolescents not to partake in these types of dangerous activities is by saying "you will kill brain cells." In this chapter we are going to try and create some clarity on the effects of drugs and alcohol on teenagers.

Both in and away from school, teens are told horror stories about drugs and alcohol. It can get confusing which are exaggerated and which are unembellished. This chapter will address these topics, and hopefully remove the shroud surrounding them so that everyone has a better understanding.

It is proven that alcohol and certain drugs impact adolescent brains far more negatively than adult brains due to brain maturation and growth happening during the teenage years. As we discussed in Chapter 2, adolescents' frontal lobes are not fully developed, myelination has not been completed, and the hippocampus has not yet fully matured. These three brain features heavily contribute to decision-making skills, therefore, teenagers tend to make slightly less rational decisions.

Alcohol
Alcohol is one of the most widely used and abused mind-altering substance: in 2013, an estimated 22.1% of high school seniors admitted to binge drinking.[lii] Many people like the way they feel when they consume alcohol, but few know what is actually happening within their brain.

When you consume an alcoholic drink, two structures are affected which create the experience of intoxication. The nucleus accumbens, the brain's reward center, produces the sense of euphoria that is commonly associated with inebriation. In adolescents, the nucleus accumbens tends to prioritize stimuli that require little effort but are very exciting or pleasurable; alcohol is a perfect match for this tendency, which is one of the reasons adolescents are more vulnerable than adults to excessive drinking .

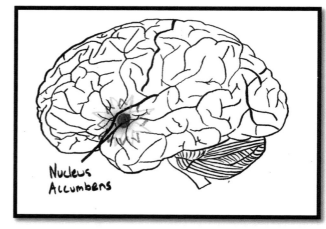

The amygdala, a part of the brain related to emotions and fear, controls emotional reactions to both pleasurable and negative experiences. In adolescents, alcohol causes the amygdala to be more inclined towards explosive reactions than towards controlled ones. This can intensify emotions and experiences. For instance, making the fact there is no available ice cream an absolute travesty and causing excessive sadness. It can become dangerous as it worsens the already compromised decision-making skills of teenagers.

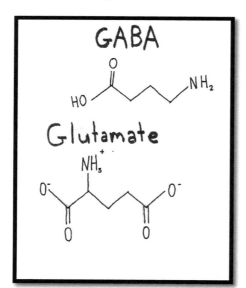

Additionally, alcohol affects neurotransmitters. The action of the neurotransmitter GABA, which is a major inhibitory neurotransmitter, is increased by alcohol. This leads to sluggishness and exhaustion, as well as depressed bodily functions (such as digestion). Glutamate, an excitatory neurotransmitter, is effected by alcohol as well. Alcohol inhibits the function of this neurotransmitter, and this is thought to be the initial cause of memory impairment during intoxication (glutamate is essential for memory formation). All of these neurobiological effects work together to increase the possibility of dangerous situations; for instance, being too sluggish to get out of the way of a car when crossing a road.

Alcohol also acts as a depressant of the Central Nervous System (CNS), slowing down nerve impulses and reaction times. It stimulates the release of endorphins, which makes you feel happy. The association this creates between alcohol and happiness, understandably, leads you to want to drink more.

When you consume alcohol, you experience reduced inhibitions, reduced coordination, drowsiness, euphoria, impaired cognition, and impaired motor skills. Adolescents experience less nausea, vomiting, and stomach pain (symptoms of excessive alcohol consumption) than adults, and as a result tend to drink more. This is especially worrying since alcohol is most damaging to teens.

The list of negative effects alcohol has on adolescents is a long one (of course, it has to be noted, that these effects are caused by frequent and excessive alcohol consumption):

1. White matter quality, which affects transmission speeds of nerve impulses, is decreased.[lxiii]
2. Exitotoxicity occurs. This is when nerve cells are excessively stimulated by neurotransmitters, and eventually cell death occurs. (This is the famed "alcohol kills brain cells").
3. Hippocampal volumes, over a long period of excessive drinking, may be reduced by up to 10%.[lxiv]
4. Alcohol consumption can cause structural lesions, which are damages to structures in the brain.
5. Binge drinkers (binge drinking is defined as consuming, in the span of a few hours, 5 or more drinks for men, and 4 or more drinks for women) have less utilization of their working memory, which is often viewed as a key determinant of intelligence. They tend to have lower verbal learning and processing skills; an effect especially prominent in adolescents, because the younger you are, the more alcohol effects memory retention.[lxv]
6. People who binge drink perform worse on thinking and memory test. Girls, specifically, perform worse on spatial functioning, which includes mathematics, and boys perform worse on tasks requiring prolonged attention.[lxvi]

7. Heavy drinkers have lower full-scale IQ scores.[lxvii]

8. Drinking can disrupt the sleep-wake cycle.

9. Depression is commonly associated with alcohol use and abuse. There are two theories for this: in extremely simple terms, either you drink because you're sad, or you're sad because you drink.

Unfortunately, most adolescents will drink before the age of 21, so while it's illegal, it's important to understand how to drink as safely as possible. An adult woman should not drink more than 2-3 units per day, no more than 14 units per week, and must have at least 2-3 alcohol-free days per week. An adult man should not drink more than 3-4 units per day, no more than 21 units per week, and must have at least 2-3 alcohol-free days per week. This is the absolute maximum your liver can process

Marijuana: The Most Commonly Used Illicit Drug

Marijuana, otherwise known as Cannabis, is a smokeable plant that contains a chemical called THC (tetrahydrocannabinol) which has psychoactive effects on the human brain. When marijuana is smoked, the THC is absorbed into the bloodstream through the lungs. It is transported to the brain, where it interacts with chemical receptors.

THC acts on the endocannabinoid system, which is a brain system that is involved with appetite, pain, mood, and memory. These, as you'll see, are areas of brain function that marijuana usage strongly impacts.

The effect most commonly associated with marijuana use is a sense of euphoria. Indeed, THC stimulates the release of dopamine, which in turn causes this feeling of happiness. However, THC also interacts with receptors in the hippocampus, amygdala, nucleus accumbens, hypothalamus, basal ganglia, and cerebellum. Respectively, these interactions impair memory, improve emotions and relieve anxiety, cause positive feelings, induce hunger, impair movement, and deteriorate muscle coordination.

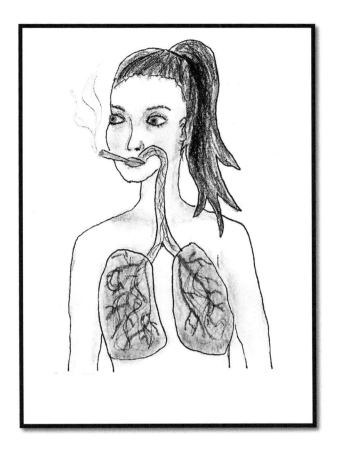

These effects are all temporary. However, cognitive function in frequent marijuana users has been shown to be different from control groups. Studies have indicated that frequent marijuana smokers perform worse on tests of cognitive flexibility, visual scanning, error commission, and working memory.[lxviii] Interestingly, however, one study showed that marijuana users had 4% *larger* posterior and prefrontal cortex volumes than non-users.[lxix] Frequent users also showed larger grey matter volumes.[lxx] During inhibition tasks, marijuana users showed the same results as controls, but used larger amounts of brain processing power in order accomplish these tasks.[lxxi] This means, essentially, that frequent marijuana users have to make more effort to control themselves.

Marijuana use during adolescence has been demonstrated to increase the risk of developing a mental disorder. Adolescents who try marijuana are 40% more likely to have psychosis (a mental disorder in which thought and emotions are severely impaired) later on in life than those who do not[lxxii], and those who use it frequently are between 50-200% as likely.[lxxiii] Additionally, excessive marijuana use can lead to amotivational syndrome, which directly leads to decline in academic performance because students with this syndrome no longer care about success.

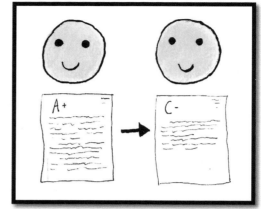

Although marijuana is only one of many illicit substances used recreationally by adolescents, it is by far the most popular. However, another class of mind-altering substance is a problem amongst the adolescent population as well: prescription drugs.

Prescription Drugs

Prescription drugs, as is indicated by the term "prescription," are medications that should only be used when, and how, they are directed by doctors. However, adolescents often take these medications recreationally to induce psychological effects. In 2004, 8.8% of teenagers ages 12-17 reported using prescription drugs for recreational – not medical – purposes.[lxxiv]

There are three classes of prescription medication that are most often abused: painkillers, sleep medications, and stimulants.

Prescription painkillers usually work by attaching to opioid receptors (these are the brain's messengers that carry information about pain). When these medications are used over long periods of times, a serious physical dependency can build up. This addiction holds the potential to wreak havoc on the lives of those who are chemically dependent upon painkillers. The painkillers can also impact the respiratory system, which has been known to cause death.

Sleep medications usually work by increasing the brain's levels of the neurotransmitter GABA (the same one that alcohol affects), which creates sleepiness. The danger with this type of medication is that if it is combined with alcohol, pain medication, or cold medication, serious side effects can occur which frequently include death, most often due to respiratory failure.

Prescription stimulant medications are thought to cause the release of dopamine and norepinephrine. While dopamine is responsible for the sensation of pleasure and reward reinforcement, norepinephrine may aid attention and focus. These kinds of medications are often prescribed for ADD/ADHD. Unfortunately, they are easily abused by students who have a lot of work to do, and some even use them recreationally, taking advantage of the pleasurable effects. As with any of these medications, prescription stimulants are extremely potent compounds that are dangerous to those who do not need them, as the dopamine stimulation can lead to addiction, and high dosages can cause serious circulatory problems.

In Conclusion
Drugs, alcohol, and prescription medication seriously affect the brain, especially in adolescents. These substances are all easily abused, and can be extremely damaging. This is worsened by adolescents' natural tendency towards risk taking and reward conditioning. Teenagers often report turning to substances in order to improve how they feel. However, fortunately, there are other safe and easy ways to positively affect how your brain works. In the next chapter, we'll explore some quick brain tricks.

References
Boyd, C. J., McCabe, S. E., Cranford, J. A., & Young, A. (2007). Prescription drug abuse and diversion among adolescents in a southeast Michigan school district. *Archives of Pediatrics and Adolescent Medicine, 161*(3), 276-281. Retrieved September 22, 2013, from http://dx.doi.org/10.1001/archpedi.161.3.276
Freudenrich, C. (2000, December 21). How alcohol works. *HowStuffWorks*. Retrieved September 24, 2013, from http://science.howstuffworks.com/alcohol6.htm

Gottlieb, E. (2012, August). Cannabis: a danger to the adolescent brain - how pediatricians can address marijuana use. *MCPAP*. Retrieved September 24, 2013, from http://www.mcpap.com/pdf/Cannibis.pdf [PDF form]

Hampson, A. J., Grimaldi, M., Lolic, M., Wink, D., Rosenthal, R., & Axelrod, J. (2000). Neuroprotective Antioxidants From Marijuanaa. *Annals of the New York Academy of Sciences*, *899*(1), 274-282. Retrieved September 30, 2013, from http://dx.doi.org/10.1111/j.1749-6632.2000.tb06193.x

Kahn, L., Kinchen, S., Shanklin, S. L., et al. Youth risk behavior surveillance – United States, 2013. MMWR 2014;63(No. SS-4):18-19.

McCance-Katz, E. F. (n.d.). Effect of drugs and alcohol on the adolescent brain. *http://sfc.virginia.gov*. Retrieved September 22, 2013, from http://sfc.virginia.gov/pdf/health/McCance-Katz%20-%20VCU%20-%20Effect%20of%20Drugs%20and%20Alcohol%20on%20the%20Adolescent.pdf [PDF form]

National Institute on Drug Abuse. (2011). Research report series: prescription drug abuse. Retrieved June 20, 2014 from http://www.drugabuse.gov/sites/default/files/rxreportfinalprint.pdf [PDF form]

Squeglia, L. M., Jacobus, J., & Tapert, S. F. (2009). The influence of substance use on adolescent brain development. *Clinical EEG and Neuroscience*, *40*(1), 31-38. Retrieved September 22, 2013, from http://dx.doi.org/10.1177/155005940904000110

Trudeau, M. (2010, January 25). Teen drinking may cause irreversible brain damage. *NPR*. Retrieved September 22, 2013, from http://www.npr.org/templates/story/story.php?storyId=122765890

Types of prescription drugs. (n.d.). *NIDA for Teens*. Retrieved September 22, 2013, from http://teens.drugabuse.gov/educators/curricula-and-lesson-plans/mind-over-matter/prescription-drug-abuse/types-prescription

Zeigler, D. W., Wang, C. C., Yoast, R. A., Dickinson, B. D., McCaffree, M. A., Robinowitz, C. B., et al. (2005). The neurocognitive effects of alcohol on adolescents and college students. *Preventive Medicine*, *40*(1), 23-32. Retrieved September 22, 2013, from http://dx.doi.org/10.1016/j.ypmed.2004.04.044

[lxii] Kahn, L., Kinchen, S., Shanklin, S. L., et al. Youth risk behavior surveillance – United States, 2013. MMWR 2014;63(No. SS-4):18-19.

[lxiii] Squeglia, L. M., Jacobus, J., & Tapert, S. F. (2009). The influence of substance use on adolescent brain development. *Clinical EEG and Neuroscience*, *40*(1), 31-38. Retrieved September 22, 2013, from http://dx.doi.org/10.1177/155005940904000110

[lxiv] Zeigler, D. W., Wang, C. C., Yoast, R. A., Dickinson, B. D., McCaffree, M. A., Robinowitz, C. B., et al. (2005). The neurocognitive effects of alcohol on adolescents and college students. *Preventive Medicine*, *40*(1), 23-32. Retrieved September 22, 2013, from http://dx.doi.org/10.1016/j.ypmed.2004.04.044

[lxv] Squeglia, L. M., Jacobus, J., & Tapert, S. F. (2009). The influence of substance use on adolescent brain development. *Clinical EEG and Neuroscience*, *40*(1), 31-38. Retrieved September 22, 2013, from http://dx.doi.org/10.1177/155005940904000110

[lxvi] Trudeau, M. (2010, January 25). Teen drinking may cause irreversible brain damage. *NPR*. Retrieved September 22, 2013, from http://www.npr.org/templates/story/story.php?storyId=122765890

[lxvii] Zeigler, D. W., Wang, C. C., Yoast, R. A., Dickinson, B. D., McCaffree, M. A., Robinowitz, C. B., et al. (2005). The neurocognitive effects of alcohol on adolescents and college students. *Preventive Medicine*, *40*(1), 23-32. Retrieved September 22, 2013, from http://dx.doi.org/10.1016/j.ypmed.2004.04.044

[lxviii] Squeglia, L. M., Jacobus, J., & Tapert, S. F. (2009). The influence of substance use on adolescent brain development. *Clinical EEG and Neuroscience*, *40*(1), 31-38. Retrieved September 22, 2013, from http://dx.doi.org/10.1177/155005940904000110

[lxix] Ibid.

[lxx] Ibid.

[lxxi] Ibid.

[lxxii] Gottlieb, E. (2012, August). Cannabis: a danger to the adolescent brain - how pediatricians can address marijuana use. *MCPAP*. Retrieved September 24, 2013, from http://www.mcpap.com/pdf/Cannibis.pdf [PDF form]

[lxxiii] Ibid.

[lxxiv] Boyd, C. J., McCabe, S. E., Cranford, J. A., & Young, A. (2007). Prescription drug abuse and diversion among adolescents in a southeast Michigan school district. *Archives of Pediatrics and Adolescent Medicine*, *161*(3), 276-281. Retrieved September 22, 2013, from http://dx.doi.org/10.1001/archpedi.161.3.276

Easy Ways to Help

The continuing research on neuroscience yields benefits for everyone. Not only does it improve medical treatment and the understanding of disorders, it also uncovers ways that you, personally, can boost your cognitive function, and improve how you feel. Many of these things are not definite solutions; some of them work for certain people, others don't. The general consensus, however, is that the following things are useful ways to boost your brain:

State-Dependent Memory

School requires huge quantities of memorization and habit-learning. Everyone has different learning methods that work well for them, but new research has shed light on an easy way to increase information retention. It has been demonstrated that recall is improved when the environment in which you're trying to remember something is the same as the one in which you learned it.[lxv] This can be applied directly to school. If you are studying vocabulary for a foreign language test, don't study it while lying down on your bed! Study it while you're sitting at a desk and focused. You are more likely to remember the vocabulary if the "state" that you learn it in is similar to the "state" in which you'll have to recall it (taking a test).

Some Helpful Tips for Sleep

Here are a few helpful tidbits that have to do with sleep. As we discussed in the chapter focused on sleep, using electronics before bed that are bright and emit blue light reduces melatonin, making it harder to fall asleep. Ideally, stop using electronics close to the time you are planning on sleeping. As this is not always possible, when you use a computer or a television at night, try and turn down the brightness. It will help you fall asleep sooner, making you more alert the next day. Another helpful thing that we discussed in the sleep chapter is getting to bed earlier. More sleep has been repeatedly correlated with improvements in academic performance, as well as increased happiness. Even though at the time, watching a TV show might seem like a sure-fire way to end the night on a high note, doing this late at night repeatedly will just build up a sleep deficit. You are *much* better off getting more sleep, and being happier the next day.

What You Eat Matters

1. <u>Eating breakfast</u>. Eating breakfast is absolutely essential. Skipping out on breakfast leaves your brain starved for energy throughout the day. Focus and information retention are impaired, making learning much more difficult. Having a breakfast with complex carbohydrates, proteins, and fats will bring very tangible changes to how you feel in the mornings.

2. <u>Proteins, Fats, and Carbohydrates</u>. You need these three substances in your diet in order for your brain to function properly. Protein helps grow new neural pathways, and leads to general learning improvements. Additionally, fats are key for myelination, which over time leads to faster and more accurate information processing. What is especially important, however, is to eat complex carbohydrates instead of refined carbohydrates. Combined with protein, which prevents blood glucose levels from dipping, complex carbohydrates leave the brain with a stable source of energy for much longer than refined carbs do. This is very important when you're in a learning environment.

Yoga and Meditation

Here's a fun one: yoga and meditation, as it turns out, can really impact your brain positively. This is what they do:

1. <u>Meditation</u>. Something known as mindfulness meditation has been shown to help the human brain. This type of meditation has been demonstrated in scientific studies to increase the sense

of well-being in participants who meditated regularly.[lxxvi] Spending a few minutes a day engaging in this type of focused relaxation could help you feel better in general – and this can help clear your mind and help you focus.

2. Yoga. Yoga, too, has some very interesting and noteworthy effects on the brain. First of all, a recent study showed that people who do yoga regularly have lower amounts of pain-related brain activity during pain stimuli than those who don't practice yoga.[lxxvii] This indicates that yoga might potentially be helpful in managing unpleasant experiences, such as chronic pain or excessive stress. In fact, one type of controlled breathing called Sudarshan Kriya yoga causes levels of the stress hormones cortisol and corticotrophin to decrease, and has also been demonstrated to help relieve some symptoms of depression.[lxxviii] This type of body movement and focusing technique has some incredible effects on the human brain, and it is well worth a shot if you struggle coping with stress.

Don't Ditch Gym Class
The mantra that exercise is good for you is cliché. But it truly is. Not just for your heart, and your muscles, and your bones, but for your brain as well. Exercising seems to do at least two things for the brain: create happiness, and help improve learning abilities. Within 30 minutes of exercising chemicals known as endorphins are released in your brain. For reasons that are not entirely known, endorphins cause mild euphoria. As interesting as this is, it is also practical. Having a bad day? Why not go for a run! Exercising is a quick, healthy way to boost your mood, something we all can benefit from.

Exercising doesn't just make you feel better; it makes your brain work better too. Humans possess a certain gene called BDNF

(Brain-Derived Neurotrophic Factor). Essentially, when activated, this gene stimulates neuronal growth. And guess what? Exercise stimulates the expression of this gene. Aerobic exercise has been shown to increase expression of BDNF genes in your brain, and therefore fosters a larger number of dendrite connections between the neurons in your brain.[lxix] This means that your neural network can process and store information better, ultimately making learning an easier task.

In Conclusion
Everything talked about in this chapter is a simple, effective way to benefit your brain. There are many other little things that can help your brain as well (did you know that chewing gum briefly improves nearly all aspects of cognitive function?). Different things work for different people, so the best way to figure out what works for you is to give them a try! Try yoga once, and see if it helps you focus or feel good. Start eating breakfast, and watch the effects it has on you. Eventually, you will find some small thing that will make a world of difference. We hope this book has provided you with some insight into the fascinating workings of your brain, and how it interacts with the world around it.

References

Beilock, S. (2011, June 3). How mindfulness meditation alters the brain. *Psychology Today*. Retrieved October 12, 2013, from http://www.psychologytoday.com/blog/choke/201106/how-mindfulness-meditation-alters-the-brain

Context and state dependent memory. (2013, March 11). *Science of Education 2013*. Retrieved October 29, 2013, from http://scienceofeducation2013.wordpress.com/2013/03/11/context-and-state-dependent-memory-blog-6-week-7-11th-march/

Lehrer, J. (2011, November 29). The cognitive benefits of chewing gum. *Wired*. Retrieved October 29, 2013, from http://www.wired.com/wiredscience/2011/11/the-cognitive-benefits-of-chewing-gum/

Mattson, M. P., Duan, W., Wan, R., & Guo, Z. (2004). Prophylactic activation of neuroprotective stress response pathways by dietary and behavioral manipulations. *NeuroRx, 1*(1), 111-116. Retrieved October 29, 2013, from http://dx.doi.org/10.1602/neurorx.1.1.111

McGovern, M. (2005). The effects of exercise on the brain. *Serendip Studio*. Retrieved October 14, 2013, from http://serendip.brynmawr.edu/bb/neuro/neuro05/web2/mmcgovern.html

University of Cambridge (2010, September 1). Mindfulness meditation increases well-being in adolescent boys, study finds. *ScienceDaily*. Retrieved October 12, 2013, from http://www.sciencedaily.com/releases/2010/09/100901111720.htm

Yoga for anxiety and depression. (2009, April). *Harvard Health Publications*. Retrieved October 12, 2013, from http://www.health.harvard.edu/newsletters/Harvard_Mental_Health_Letter/2009/April/Yoga-for-anxiety-and-depression

Avery Bedows

[lxxv] Context and state dependent memory. (2013, March 11). *Science of Education 2013*. Retrieved October 29, 2013, from http://scienceofeducation2013.wordpress.com/2013/03/11/context-and-state-dependent-memory-blog-6-week-7-11th-march/

[lxxvi] University of Cambridge (2010, September 1). Mindfulness meditation increases well-being in adolescent boys, study finds. *ScienceDaily*. Retrieved October 12, 2013, from http://www.sciencedaily.com/releases/2010/09/100901111720.htm

[lxxvii] Yoga for anxiety and depression. (2009, April). *Harvard Health Publications*. Retrieved October 12, 2013, from http://www.health.harvard.edu/newsletters/Harvard_Mental_Health_Letter/2009/April/Yoga-for-anxiety-and-depression

[lxxviii] Ibid.

[lxxix] McGovern, M. (2005). The effects of exercise on the brain. *Serendip Studio*. Retrieved October 14, 2013, from http://serendip.brynmawr.edu/bb/neuro/neuro05/web2/mmcgovern.html

Made in the USA
Charleston, SC
12 September 2014